Seventh International Workshop on
MICROPROCESSOR TEST AND VERIFICATION
MTV 2006

PROCEEDINGS

· ·

AA001080

Seventh International Workshop on Microprocessor Test and Verification (MTV 2006)

**Austin, Texas
4-5 December 2006**

IEEE Catalog Number: CFP06MTV-POD
ISBN: 978-0-76952-839-7

Seventh International Workshop on
MICROPROCESSOR TEST AND VERIFICATION
MTV 2006

PROCEEDINGS

Austin, Texas
4-5 December 2006

Sponsored by
IEEE Computer Society
Test Technology Technical Council (TTTC)

Los Alamitos, California

Washington • Tokyo

Copyright © 2007 by The Institute of Electrical and Electronics Engineers, Inc.
All rights reserved.

Copyright and Reprint Permissions: Abstracting is permitted with credit to the source. Libraries may photocopy beyond the limits of US copyright law, for private use of patrons, those articles in this volume that carry a code at the bottom of the first page, provided that the per-copy fee indicated in the code is paid through the Copyright Clearance Center, 222 Rosewood Drive, Danvers, MA 01923.

Other copying, reprint, or republication requests should be addressed to: IEEE Copyrights Manager, IEEE Service Center, 445 Hoes Lane, P.O. Box 133, Piscataway, NJ 08855-1331.

The papers in this book comprise the proceedings of the meeting mentioned on the cover and title page. They reflect the authors' opinions and, in the interests of timely dissemination, are published as presented and without change. Their inclusion in this publication does not necessarily constitute endorsement by the editors, the IEEE Computer Society, or the Institute of Electrical and Electronics Engineers, Inc.

IEEE Computer Society Order Number P2839
ISBN-10 0-7695-2839-2
ISBN-13 978-0-7695-2839-7
ISSN Number 1550-4093

Additional copies may be ordered from:

IEEE Computer Society	IEEE Service Center	IEEE Computer Society
Customer Service Center	445 Hoes Lane	Asia/Pacific Office
10662 Los Vaqueros Circle	P.O. Box 1331	Watanabe Bldg., 1-4-2
P.O. Box 3014	Piscataway, NJ 08855-1331	Minami-Aoyama
Los Alamitos, CA 90720-1314	Tel: + 1 732 981 0060	Minato-ku, Tokyo 107-0062
Tel: + 1 800 272 6657	Fax: + 1 732 981 9667	JAPAN
Fax: + 1 714 821 4641	http://shop.ieee.org/store/	Tel: + 81 3 3408 3118
http://computer.org/cspress	customer-service@ieee.org	Fax: + 81 3 3408 3553
csbooks@computer.org		tokyo.ofc@computer.org

Individual paper REPRINTS may be ordered at: <reprints@computer.org>

Editorial production by Silvia Ceballos
Cover art production by Joe Daigle/Studio Productions
Printed in the United States of America by Applied Digital Imaging

Conference Publishing Services
http://www.computer.org/proceedings/

Seventh International Workshop on
MICROPROCESSOR TEST AND VERIFICATION
MTV 2006

TABLE OF CONTENTS

Preface vii

Acknowledgement viii

Workshop Organizing Committee ix

Program Committee x

SECTION 1: TEST

Software-Based On-Line Test of Communication Peripherals in Processor-Based Systems
for Automotive Applications 3
    P. Bernardi, L. Bolzani, A. Manzone, M. Osella, M. Violante, and M. Sonza Reorda

Circuit Profiling Mechanisms for High-Level ATPG 9
    Jorge Campos and Hussain Al-Asaad

Functional Test Selection for High Volume Manufacturing 15
    Vijay Gangaram, Deepa Bhan, and James K. Caldwell

Test Calculation for Logic and Delay Faults in Digital Circuits 20
    József Sziray

SECTION 2: VERIFICATION AND TEST GENERATION

Directed Micro-architectural Test Generation for an Industrial Processor: A Case Study 33
    Heon-Mo Koo, Prabhat Mishra, Jayanta Bhadra, and Magdy Abadir

Advanced SAT-Techniques for Bounded Model Checking of Blackbox Designs 37
    Marc Herbstritt, Bernd Becker, and Christoph Scholl

Embedded Software Validation: Applying Formal Techniques for Coverage and Test Generation 45
    Tamarah Arons, Elad Elster, Terry Murphy, and Eli Singerman

Challenges in System on Chip Verification 52
    Noah Bamford, Rekha K Bangalore, Eric Chapman, Hector Chavez,
    Rajeev Dasari, Yinfang Lin, and Edgar Jimenez

SECTION 3: ARCHITECTURAL AND DESIGN ISSUES

Workload Slicing for Characterizing New Features in High
Performance Microprocessors 61
    Hassan Al-Sukhni, David Lindberg, James Holt, and Michele Reese

Deep vs. Shallow, Kernel vs. Language – What is Better for Heterogeneous Modeling in SystemC? 68
    Hiren D. Patel and Sandeep K. Shukla

Statistical Static Timing Analysis Considering the Impact of Power Supply Noise in VLSI Circuits 76
    Hyun Sung Kim and D. M. H. Walker

v

SECTION 4: DESIGN ERROR DEBUG & DIAGNOSIS

Debug Support for Scalable System-on-Chip 83
  Jianmin Zhang, Ming Yan, and Sikun Li

Abstraction and Refinement Techniques in Automated Design Debugging 88
  Sean Safarpour and Andreas Veneris

Diagnosing Silicon Failures Based on Functional Test Patterns 94
  Chia-Chih Yen, Ten Lin, Hermes Lin, Kai Yang, Tayung Liu, and Yu-Chin Hsu

AUTHOR INDEX 99

PREFACE

The papers presented in this book have been revised from the original submissions to the Seventh IEEE International workshop on Microprocessor Test and Verification (MTV), held in Austin, Texas in December 2006 and sponsored by IEEE Computer Society Test Technology Technical Council (TTTC).

The topic area, applications of verification, validation and test to complex electronic circuits at all levels, has blossomed considerably since the first workshop was held in 1999. The scope of the workshop has expanded beyond just microprocessors to include all types of complex integrated circuits and Systems-on-Chip (SOCs). The 2006 workshop was certainly the most successful one of the series so far, and we would like to thank all participants who contributed to this event.

High level functional verification remains a key challenge facing designers of complex SOCs and microprocessors. This is reflected in large number of papers in this year's MTV. These papers discuss issues related to architecture description languages, validation, sequential equivalence checking, high level test generation, design debug/diagnosis. Functional verification and ATPG is one synergetic area that has evolved significantly in recent years due to the blossoming of a wide array of test and verification techniques. This area will continue to be a key focus of future MTV events. This year's MTV had a panel with a related topic – "Strategies for Convergence between Design Validation, Test and Debug".

To encourage industrial experts to openly discuss the current practice, we did not request written paper publication for every presentation. Hence, this proceeding includes only the selected papers contributed by the authors and presenters. We will continue to adopt this strategy in order to encourage industrial participants to share their experience and results via the MTV forum. Interested readers who want to learn more about the current industrial practice can consult the MTV web site http://mtv.ece.ucsb.edu/MTV/ for future events.

Magdy S. Abadir, General Chair

Li-C. Wang, Program Chair

Jayanta Bhadra, Program co-Chair

ACKNOWLEDGMENT

Many people contributed to the success of MTV and to the publishing of this proceeding. We thank all the contributors for their sustained interest and diligence. We are deeply indebted to all members of the organizing and program committees for their support to MTV events over the years.

We also would like to thank our sponsors, the Test Technology Technical Council and the IEEE Computer Society.

Many thanks to IBM Haifa, Pintail Technologies, Obsidian and Freescale Semiconductor for their continued support to MTV and for all technical and monetary contributions.

Special thanks to Heather Bethea for helping with various logistics during MTV.

WORKSHOP ORGANIZING COMMITTEE

GENERAL CHAIR
Magdy S. Abadir (m.abadir@freescale.com), *Freescale Semiconductor*

PROGRAM CHAIR
Li-C. Wang (licwang@ece.ucsb.edu), *University of California at Santa Barbara*

PROGRAM CO-CHAIR
Jayanta Bhadra (jayanta.bhadra@freescale.com), *Freescale Semiconductor*

ORGANIZING COMMITTEE

FINANCE
Magdy S. Abadir (m.abadir@freescale.com), *Freescale Semiconductor*

PUBLICATION
Alper Sen, *Freescale*

PANEL
Al Crouch, *Inovys*

PUBLICITY
Tao Feng, *Cadence*

COMMITTEE
Moshe Levinger, *IBM, Israel*
Jennifer Dworak, *Brown University*

EUROPEAN/CANADIAN
Andreas Veneris, *University of Toronto, Canada*

PROGRAM COMMITTEE

Jacob Abraham, *UT-Austin*
Miron Abramovici, *DAFCA*
Hussain Al-Asaad, *UC-Davis*
Tony Ambler, *UT-Austin*
Eyal Bin, *IBM - Haifa*
Shawn Blanton, *CMU*
Melvin Breuer, *USC*
Ken Butler, *TI*
Yiring-An Chen, Synopsys
K.-T. (Tim) Cheng, *UCSB*
Nick Dutt, *UC-Irvine*
Sujit Dey, *UC- San Diego*
Ajit Dingankar, *Intel*
Franco Fummi, *Universita `di Verona*
Mike Garcia, *Freescale*
Sandeep Gupta, *USC*
Ian Harris, *U. Mass.*
John Hayes, *U. Michigan*
Eric Hennenhofer, *Obsidian, Inc.*
Alan J.Hu, *UBC, Canada*
T.M.Mak, *Intel*
Anmol Mathur, *Calypto*
Hillel Miller, *Freescale*
Sankaran Menon, *Intel*
Ishwar Parulkar, *Sun*
Carl Pixley, *Synopsys*
Paolo Prinetto, *Poli di Torino*
Nur Touba, *UT-Austin*
Miroslav Velev, *CMU*
Vivekananda Vedula, *Intel*
Cheng-Wen Wu, *National Tsing-Hua University*
Paul R Zehr, *Intel*
Yervant Zorian, *VirageLogic*

SECTION 1: TEST

Software-based on-line test of communication peripherals in processor-based systems for automotive applications

P. Bernardi**, L. Bolzani**, A. Manzone*, M. Osella*, M. Violante**, M. Sonza Reorda**

Centro Ricerche Fiat, Torino, Italy
**Politecnico di Torino – Dip. di Automatica e Informatica, Torino, Italy*

ABSTRACT

The adoption of Systems-on-a-Chip (SoCs) in automotive systems opens interesting possibilities, but also introduces significant dependability concerns. Up to now, researchers focused most of their efforts in devising new solutions for improving the dependability of the processor-cores embedded in typical SoCs, and several solutions mixing software techniques with hardware ones have been proposed, which result in low-cost dependable systems. Conversely, the peripheral components also typically embedded in SoCs are often designed according to traditional area-demanding hardware-only fault tolerance techniques.

In this paper, we propose an experimental evaluation of the effectiveness of a purely software-based approach, which can be easily and inexpensively implemented on existing SoCs. We present results on a case study inspired to a real-life application, which exploits a network of SoCs based on the Motorola 6809 processor core: experiments show that the approach achieves relatively high fault coverage with relatively reduced performance penalties.

1. INTRODUCTION

The development of vehicle electronic systems, due to the continuous improvement of vehicle performance, requires meeting ambitious goals from the performance, reliability, safety, and functionality points of view. This problem is exacerbated when we consider the integration of several subsystems (e.g., *Vehicle Dynamics Control* or VDC, *Anti-blocking Brake System* or ABS) that need to collaborate to provide integrated vehicle functions.

Currently, car manufacturers are facing a radical shift in their design methodologies: while relying on different suppliers to provide Electronic Control Units (ECUs) implementing control strategies, each high-level vehicle function (e.g., trajectory control) depends on the integrated interaction of many ECUs, and each ECU contributes to the control of many vehicle functions. The role of the system integrator, i.e., the car manufacturer, therefore includes the definition of the general vehicle strategies, their partitioning in sub-functions that may be allocated to different ECUs, and the outsourcing of each ECU to the selected supplier, with a detailed-enough specification to guarantee correct integration of the related sub-function in the vehicle. These modeling and specification tasks are further worsened by the need of devising comprehensive diagnostic and recovery strategies (that in real ECUs take most of the code and the development effort) in case transient or permanent faults arise in the system [1].

Integrated Safety Systems development takes care of the definition of the reliability requirements for the electronic parts in the automotive environments. For the realization of such Integrated Safety Systems a powerful and highly dependable in-vehicle electronic architecture and an appropriate development support are mandatory. Those elements, which are relevant for both Original Equipment Manufacturers (OEMs) and suppliers, have to be standardized to achieve an improvement of system quality at shorter development times and lower system costs. Because of the aforementioned reasons, automotive-oriented hardware architecture definition currently addresses vehicle on-board electronic hardware infrastructure that supports the requirements of integrated safety systems in a cost effective manner. By means of hardware infrastructure design, it is intended the development of a logic system networked architecture composed of several Electronic Control Units (ECUs) and a proper ECU internal hardware architecture suitable for coping with functionality and reliability requirements.

The principal requirements addressed by such hardware infrastructures are [2]:

- Scalability, for standardized usage for safety relevant and non-safety relevant applications
- Flexibility, to cope with the different application domains
- Architectural simplicity
- Capability of handling a large variety of sensors and actuators
- Support for fault tolerance, error detection and error handling
- Well defined and standardized interfaces
- Means for functional integration into silicon
- Optimized costs and reliability.

In this paper we mainly deal with the last four items. In particular, we discuss the critical aspects coming from the use of Systems-on-chip (SoCs) in automotive environments. Such highly integrated silicon components allow for high performances and reduced costs, but stigmatize the reliability problems, since they include elaboration, communication and transmission functional modules and rise the need for analyzing the fault tolerance requirements.

We mainly focus on reliability requirements for networked ECUs included in Integrated Safety Systems and we address the reliability issues of peripheral devices employed for data transfer: we take into account *permanent* faults and analyze the possible effects led by faults possibly affecting the peripheral devices. Several solutions can be adopted for guaranteeing that these kinds of faults, if arisen, are detected within a maximum latency time. Due to cost constraints, there is currently a high interest for software-based techniques, that do not require any significant change in the hardware architecture in order to be implemented, and can be easily tuned during the system software development. This interest is also supported by the increasing popularity of software-based solutions exploited for the test of the processor cores [6]. The question is now how effective these techniques are in the test of peripheral components, that are a crucial component of many SoCs.

The main contribution of the paper is to provide an experimental evaluation of the effectiveness of such techniques when applied to the test of some common communication peripherals used in real systems. In particular, we focused on a solution based on the periodical execution of suitable software-based self-test programs oriented to data exchanging between processor units included in different ECUs. The test programs are activated during the application idle times following the non-concurrent on-line test scheme. As efficiency parameters, we considered the

1550-4093/07 $25.00 © 2007 IEEE

impact of the investigated approach on the automotive system mission functionalities and we provide practical rules aimed at reducing the test procedure duration. This goal is achieved by selecting a subset of faults in the whole peripheral fault list that is worked out from analysis of the system mission behaviour.

The following paragraphs are organized as it follows. In section 2, backgrounds and requirements of the proposed work are summarized. Section 3 describes the technique we considered for the on-line test of permanent faults involving the communication protocol, while section 4 proposes the results obtained on a case study based on a Motorola 6809 processor. Section 5 underlines the most critical problems encountered and the obtained results, thus providing a set of conclusions.

2. PREVIOUS WORK

Testing SoCs is a well-known problem that involves both the accessibility to each single functional module, or core, and the test application frequency. The effort of test engineers usually concentrates on the definition of test techniques and structures, which allow the efficient test of each core, even if deeply embedded.

From the point of view of on-line SoC test, several techniques have been proposed, that can be categorized as follows:

- *Codification-based approaches*: these approaches rely on the codification of the elaborated data and allow to concurrently check the correctness of the computation during device mission mode [3].
- *Hardware-based techniques*: they can be further grouped in:
  1. *Infrastructure IP-based approaches*: they consists in the insertion of additional hardware structures called Infrastructure IPs (I-IPs), suitably included in the SoC structure to concurrently check the correctness of some (possibly the greatest number) of the cores composing the system [4].
  2. *Reuse-Based approaches*: these approaches reuse the hardware structures included in the SoC for manufacturing test purposes to periodically check the controlled core functionalities [5].
- *Synergy-based techniques*: these techniques exploit the internal structure of the SoC to periodically perform self-checking operations aimed at autonomously isolate faulty cores [2][6].

Synergy-based techniques suit well in the automotive field. ECUs included in the vehicle control network are usually complex SoCs including computational units (typically one or more microcontrollers), devoted to the management of most of the car functionalities (i.e., engine, ABS, steer, etc.). Figure 1 shows a conceptual view of a generic vehicle networked electronic environment.

In a typical synergy-based strategy, a processor-core included in one ECU executes a suitable test program that implements the following operations:

1. It sends a request to another processor included in the ECU.
2. It waits the response from the polled processor, and compares the obtained response with an expected pre-computed value.
3. In case of mismatch, it switches the system to a safe mode, or starts a system recovery procedure.

When applying such a technique, engineers should consider at least the following complementary aspects:

- The test program must leave the processor configuration unchanged with respect to the original

functional configuration; conflicts between the tests of different modules should be avoided.
- A proper time slot must be identified to run the test program, which can not run concurrently with the normal operations of the system.

Two possible application strategies can be employed, depending on the system requirements and availability:

- *Event Triggered*: the test program is executed as a consequence of an event (e.g., idle clock cycles, key-on/off).
- *Time Triggered*: the test program is executed periodically.

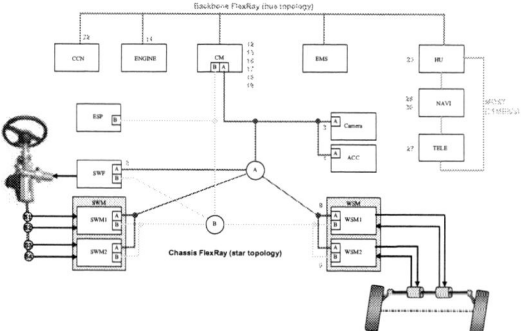

Fig 1: conceptual view of a generic electronic environment.

The characteristics of a test program suitable for on-line testing are enounced in [6] regarding processor Software-Based Self-Test (SBST) programs for on-line periodic testing. These requirements can be easily moved to synergy-based tests that must have the following stringent characteristics. A set of small test programs has to be produced, able to guarantee the highest faults coverage. Each program should be the shortest possible in terms of test execution time (less than a quantum time cycle). It should also corresponds (whenever possible) to small code without unresolved data hazards and with as much as possible compact loops that take advantage of temporal locality and sequentially executed instructions.

3. THE CONSIDERED TECHNIQUE

In this paper, we consider a software-based technique for the on-line test of peripheral devices embedded in the SoCs used in safety critical networked environments.

The considered technique is based on suitably generated test programs to be run by the microcontrollers included in SoCs and managing the inspected environment; such test programs activate the peripheral device to microcontroller data exchanging functionalities, and check for the correctness of the operation. In particular, we concentrate on communication peripheral devices.

In practical terms, the methodology exploits the existing synergies between the different SoCs included in the network, forcing the concurrent execution of test programs oriented to the test of the SoCs communication protocol. For synergy it is intended the mechanism for which a SoC included in the considered environment is able to exchange data with other SoCs in the same functional environment.

In the following paragraphs, the faulty communication behaviors to be detected and the test programs characteristics allowing a high coverage to be reached are discussed.

Figure 2 shows the conceptual view of a generic investigated environment including 4 SoCs, each of them able to exchange data through their I/O communication devices.

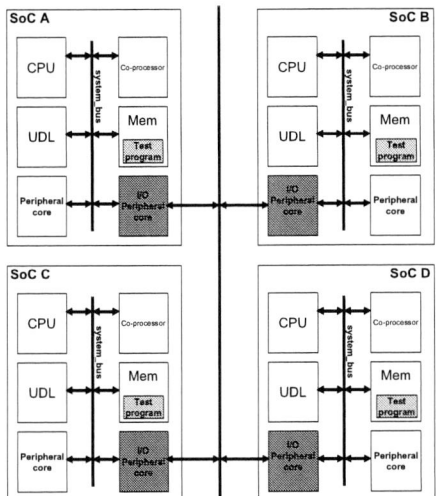

Fig 2: conceptual view of a generic electronic environment: the investigated peripheral cores are shown in red.

3.1. The considered fault models

For our purposes, we considered Stuck-At (SA) faults as the target fault model in I/O communication devices for evaluating the effectiveness of the proposed on-line test strategy.

From the user's point of view, SA faults affecting communication devices may have the following effects:

- *Early Message*: The message is received earlier than a reception time range Tmin-Tmax.
- *Faulty Data Value*: Data value is received within the reception time range Tmin-Tmax but it does not match the expected value.
- *Late Message*: The message is received later than a defined reception time range Tmin-Tmax but before a defined maximum time Ttimeout
- *Time Out*: The message is not received in a defined maximum time Ttimeout.
- *No Effect*: The data transmission is executed as if the faults were not present.
- *Detected*: The data value is not transmitted but the fault is detected by the internal detection mechanisms of the communication device or by the fault tolerance techniques added to system.

The proposed test strategy takes into account these wrong behaviors and exploits test procedures suitably generated to cope with each of these undesirable situations.

3.2 Test programs structure

Similarly to Software-Based Self-Test programs used to test processors in a functional manner, the considered technique consists in the execution of appropriate distributed test programs. Each of the involved SoCs is in charge of internally executing a suitable code able to open, evaluate and close the communication within a partner SoC; this involves the exchange of suitable data able to excite and propagate the highest number of faults.

Figure 3 graphically represents the distributed test program tasks and refers to SoC A and SoC B included in the system shown in figure 2: the processor included in SoC A sends a data to SoC B by means of executing the test program code stored in an available memory. On the contrary, the

processor in SoC B executes a test program that poses the system in reception mode. As soon as the SoC B receives a value, it is compared against a pre-calculated value and a new data is sent by the SoC B to the SoC A now holding in a waiting mode.

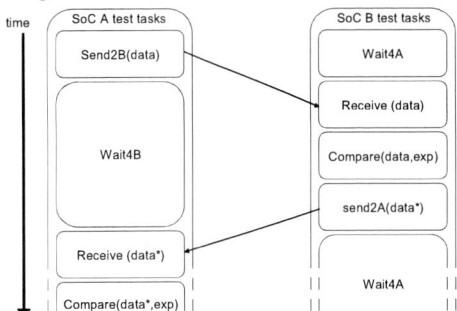

Fig 3: distributed test code synergy functionalities.

The data exchanged in the SoCs communication should be able to excite and propagate the highest percentage of the faults potentially affecting the peripheral devices involved in the transmission/reception procedure. For matching the coverage requirements, the following rules have to be followed:

1. All the functionally employed peripheral configurations have to be used in order to excite the faults involving the configuration registers and circuitries
2. The data registers and circuitries have to be stressed by data values such as Walking or Checkerboard sequences [8].

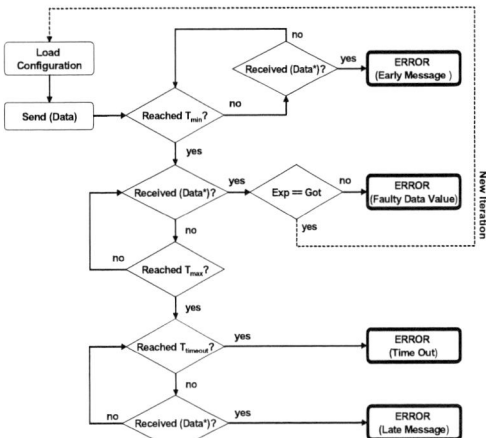

Fig 4: test program code with error detection capabilities.

The test programs have to be written in such a way that a suitable ERROR procedure is activated in the case one of the faults listed in section 3.1 occurs; the mandatory error detection functionalities to be included in the test programs are detailed in the flowchart shown in figure 4.

For the sake of on-line testing, the provided test programs should respect the requirements underlined in Section 2 for the on-line Software-Based Self-Test of processor cores. Suitable time slots have to be individuated during the environment mission mode idle times. Each test program should be the shortest possible in terms of test execution time in order to affect the availability of the system as less as possible. Each test program should also be the shortest

1550-4093/07 $25.00 © 2007 IEEE

possible in terms of memory occupation in order not to affect the memory system requirements.

Accordingly to these constraints, we defined an *atomic* test procedure composed of 2 test programs:

- A *Sender* test program executed by the processor core of a SoC included in the system executing 2 send operations and 1 receive operation
- A *Receiver* test program executed by the processor core of another SoC included in the system that performs 2 receive operations and 1 send operation.

Each of these test programs must leave the SoC state unchanged after its execution. In practical terms, the final test program structure must be the following: a preliminary instruction sequence has to be run to save the current peripheral configuration; the test program is executed and, eventually, an ERROR condition signaled; a final instruction sequence is in charge of reloading the original peripheral configuration.

The defined test procedure suits to be executed during system idle periods: depending on the system functionalities, a test procedure may be divided in several sub-procedures, whose duration is customized on the mission mode characteristics.

Finally, test code length and test duration are minimized by selecting a subset of the original peripheral fault list; this subset is obtained by considering which faults may affect the functionalities of the working system. In simple words, the defined test sequence should address only the faults in those parts of the investigated devices utilized during the mission mode. Such consideration permits saving execution time, code length and, especially, generation efforts.

4. EXPERIMENTAL RESULTS

In this section, we describe how the considered software-based on-line test technique has been evaluated and we summarize the results obtained from fault injection experiments performed on a realistic scenario consisting of a system including two SoCs equipped with Motorola 6809 processors as well as serial and parallel transmission peripherals cores.

The following paragraphs are organized as follows. In sub-section 4.1 we discuss the case study hardware structure and its mission functionalities. Sub-section 4.2 describes the process devised for selecting the set of the critical faults provoking communicational malfunctioning, while sub-section 4.3 presents the results obtained from fault injection campaigns including a detailed misbehavior classification and a example of the test programs structure developed for evaluating the effectiveness of our approach. Finally, in sub-section 4.4 we analyze the implementation costs of the proposed technique.

4.1 Description of the case study

As a realistic case study, we considered a networked system including two SoCs, each of them including a Motorola 6809 processor, a RAM memory core storing the program code, a serial peripheral device (*Universal Asynchronous Receive and Transmit*, or UART) and a parallel peripheral device (*Peripheral Interface Adapter*, or PIA) used to establish the communication between the two SoCs. Figure 5 shows the schematic view of the considered environment.

For the sake of this paper, the system we considered is intended to execute the application functionalities presented in the paper [10], where the network embedded in a realistic vehicle is presented. The network is composed of several nodes transmitting information about the status of the vehicle to a vehicle dynamic controller (VDC), whose purpose is to control that the vehicle performs any maneuver in a safe and

comfortable way. In this paper we model any node transmitting information as a generic *Sender* node, while we model the VDC as a *Receiver* node. Dedicated serial and parallel lines connect the *Sender* and *Receiver* nodes. The communication between Sender and Receiver is implemented as an open loop:

- The Sender sends a data packet every 10 ms
- The Receiver receives and processes the information coming from the Sender.

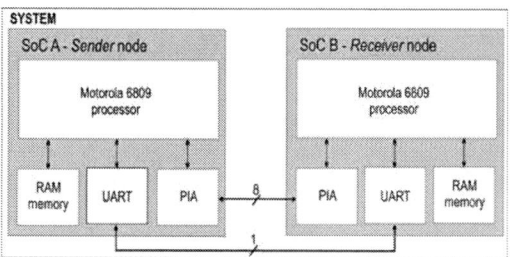

Fig 5: the experimental setup used for evaluating the proposed technique.

Considering the architecture shown in figure 5, we assume that:

- SoC A corresponds to the Sender node, which sends a data to SoC B every 10 ms
- SoC B corresponds to the Receiver node that receives the data to be elaborate from Soc A.

The Motorola 6809 processor cores included in each SoC are initialized to work in four communication modes. These modes are implemented using the following configurations:

- Case A: in this communication mode, the UART is configured by the control register, the status register is polled and the data values are sent when ready.
  - o *Sender node*: this node sends random data values to the Receiver node. The data values are transmitted in the following transmission mode: data sent without parity, 2 stop bits and a transmission rate ratio of 1;
  - o *Receiver node*: this node receives the data values that have been sent from Sender node;
- Case B: in this communication mode, the UART is also configured according to the control register value to work in interrupt mode. In this mode, when an interrupt occurs the serial device sends a new data value.
  - o *Sender node:* this node sends random data values to the Receiver node. The data values are transmitted using the same configuration used in the Case A;
  - o *Receiver node:* this node receives the data values that have been sent from Sender node;
- Case C: in this communication mode, the PIA is configured according to the control register value; the data direction register and the status register is polled during the data transmission.
  - o *Sender node*: this node sends random data values to the Receiver node by a port that has been configured as an output. The Sender node wait for an acknowledge signal from Receiver node to send the next data value;
  - o *Receiver node*: this node receives the data values and sends an acknowledge signal to the Sender node;
- Case D: in this communication mode, the PIA is configured according to the control register value and

it sends a data when an interrupt request is raised from the parallel device. An interrupt request is raised anytime the acknowledge signal is sent by the Receiver node.

o *Sender node:* this node sends random data values to the Receiver node by a port that has been configured as an output. The Sender node sends the next data value when an interrupt occurs.

o *Receiver node:* this node receives the dada values and sends an acknowledge signal to the Sender node;

4.2 Critical faults selection

When considering on-line testing issues, only the set of faults that can affect a device is relevant; as a matter of facts, some faults cannot modify the system behavior since the circuitry parts they belong to are not used during the system's mission. For this reason, we selected a set of critical faults for each communication mode described in the sub-section 4.1. In our case, the faults that we consider as critical for the UART and PIA are represented by a fault set causing a wrong communication.

In more precise words, critical faults are those faults modifying the peripheral core output in the functional mode. This selection is mainly done by running a fault simulation process considering the functional program as a set of patterns to be evaluated. In our environment we gathered the selection resources, in table 1.

Case	Total faults [#]	Critical faults [#]
A	4,333	1,473
B	4,333	1,229
C	3,892	1,423
D	3,892	1,516

Tab 1: total faults versus critical faults.

The other faults have been classified as *effect-less* with respect to the required system functionalities.

4.3 Test procedure characteristics and detection abilities

In the context described in sub-section 4.1 we wrote 2 test programs (one for each SoC) for each communication mode previously described. These test programs implement the following procedure:

- Data is exchanged between the two SoCs
- Data reception is checked and misbehavior classified recurring to the faulty classification introduced in sub-section 3.1:
  o **Faulty Data Value** error, if the expected and gotten data is different or an internal error flag signaled an error in the transmission
  o **Timeout** error, if the expected data is never received.
  o **Early Message** and **Late Message** errors are not specifically classified. In the investigated case these misbehaviors are included in the Timeout case.

The figure 6 shows a schematic example of the UART's test program developed to work in polling mode where 1 walking and 4 checkerboard data patterns are exchanged between SoC A and B. The test procedures included a comparison code part that allows checking data consistency and arrival times.

The fault coverage capability over the critical fault sets of the software-based technique proposed in sub-section 4.2 is verified by performing fault injection experiments. The table 2 shows the fault coverage capability of each test program developed specifically for each functional case mentioned in

the sub-section 4.1. These test programs present the following characteristics:

- **Test program A:** this test program configures the UART for working in the polling mode and executes repeated data transmission and elaboration. The UART has been configured in the following transmission modes:
  o data sent with odd parity bit, 1 stop bit and a transmission rate ratio of 1;
  o data sent without parity bit, 2 stop bits and a transmission rate ratio of 1;

- **Test program B:** this test program configures the UART to work in interrupt mode (transmit and receive interrupt) and is based on the same transmission principle as the test program A. The UART has been configured in the following transmission modes:
  o data sent with even parity bit, 1 stop bit and a transmission rate ratio of 1;
  o data sent without parity bit, 2 stop bits and a transmission rate ratio divided by 1;

- **Test program C:** this test program configures the PIA to work in polling mode and executes repeated data transmissions and elaboration. Each of the PIA's ports has been configured as input as well as output and the control signals use all possible function modes.

- **Test program D:** this version of the test program configures the PIA to work in interrupt mode and is based on the same transmission idea as the test program C.

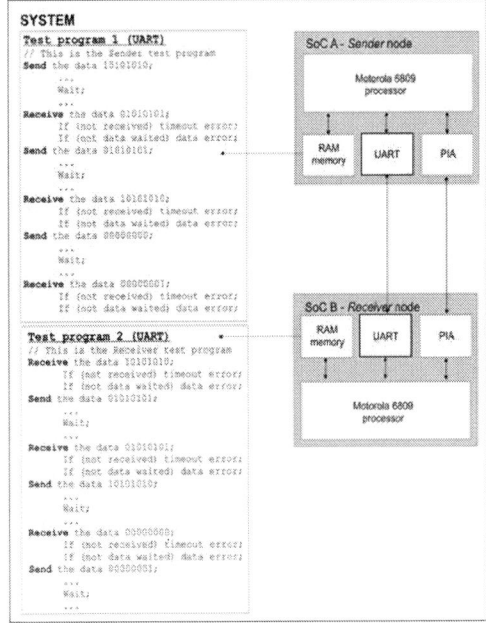

Fig 6: example of UART's test program structure based on the polling mode.

The test program A achieves 98% coverage of the total faults injected during the fault injection experiments. Most of these faults, about 75.5% are classified as data errors. In the case of the test program B the obtained fault coverage is 95.7% of the total faults injected. Also in this case the most of these faults are classified as data errors, about 56.2%.

The test program C detects 94% of the total faults injected during the fault injection experiments. Finally, the test program D achieves 91% of the total faults injected. In both cases the most faults are classified as timeout error, about 72% in case of the test program C and 73.5% for the test program D.

Analyzing the table 2 it is possible observing that in the case of the UART most of the detected faults are classified as data errors and in the case of the PIA most of the detected faults are classified as timeout errors.

4.4 Implementation costs

Implementation costs introduced by the proposed technique for on-line test of peripheral cores are mainly related to memory occupation and required execution time. The test programs have been written in the assembly language of the Motorola 6809 processor. The table 3 summarizes the area and time overhead for each test program developed.

Test program	Length [bytes]	Time [us]
A	400	300
B	336	406
C	440	90
D	452	96

Tab 3: area and time overhead.

An important consideration regards the fact that the proposed technique does not require additional circuitry especially devised for tests since our method is purely based on software procedure execution. The execution times presented in the table 3 are calculated assuming a 10 MHz system clock frequency. Anyway, the test procedure is executed during the time period between two consecutive functional data transmissions, thus does not affect the mission's execution time.

An alternative viable solution consists in exploiting some kind of hardware and time redundancy.

Hardware redundancy, commonly based on Triple Modular Redundancy (TMR), offers full fault detection and correction of the UART faults. This technique, if applied on the inspected case study, introduces at least a 200% area overhead, requesting for about 5,000 additional gates; moreover, the TMR technique (if applied to the transmission cores) is not able to check the communication bus integrity.

The Time Redundancy technique consisting in replicated transmissions and received data comparison. Such an approach guarantees the same detection capabilities achieved by the proposed method. The costs for this technique involve both the memory area occupied and the mission time. Regarding the additional memory area required, we observed for the used mission code an overhead of about 57%, due to duplicated instructions and comparison code parts. The application time is further increased to its double, reducing the transmission ratio of about a half.

5. CONCLUSIONS

The paper presents the application of software-based on-line test technique to the test of the most common peripheral devices embedded in a network of SoCs. It basically consists in the execution of test programs that are run during idle communication periods, thus not affecting the functionality of the system.

The functional tests abilities and requirements are detailed in the paper, and their effectiveness evaluated on a realistic case study coming from the automotive field. The proposed approach is valuable because it is able to identify efficiently the set of critical faults depending on the mission functionalities of the system. These faults may modify drastically the functionality of these systems and therefore designers need to identify them as soon as possible. We consider our work a major contribution to the identification of these critical faults, thus providing a starting point for the development of mechanisms capable of increasing the reliability of the peripherals on the networked systems. The proposed method allowed detecting up to 98% of the selected critical faults, introduces negligible memory area overhead and no performance reduction.

6. REFERENCES

[1] D.K. Pradhan, "Fault-Tolerant Computer System Design", Prentice Hall PTR, 1996

[2] M.Osella, A.Ferre, B.Hedenetz, D.Bugnot, D.V.Wageningen, E.Johansson, F.Camut, M.Menzel, M.Jordan, M.Sciolla, T.Söderqvist, EASIS Project, IST 2002-507690, Deliverable D2.2 "Conceptual Hardware Architecture Specification"

[3] M. Favalli, C. Metra, "Single output distributed two-rail checker with diagnosing capabilities for bus based self-checking architectures", IEEE International On-Line Testing Workshop, 2001, pp. 100 - 105

[4] Y. Zorian, "What is an Infrastructure IP?", IEEE Design & Test of Computers, Vol. 19, No. 3, May-June, 2002, pp.5-7

[5] A. Manzone, P. Bernardi, M. Grosso, M. Rebaudengo, E. Sanchez, M.Sonza Reorda, "Integrating BIST techniques for on-line SoC testing", IEEE International On-Line Testing Symposium, 2005, pp. 235 - 240

[6] A. Paschalis, D. Gizopoulos, "Effective Software-Based Self-Test Strategies for On-Line Periodic Testing of Embedded Processors", IEEE Transactions on Computer-Aided Design of Integrated Circuits and Systems, Volume: 24, Issue: 1, Jan. 2005, pp. 88 – 99

[7] K. Yamasaki, I. Suzuki, A. Kobayashi, K. Horie, Y. Kobayashi, H. Aoki, H. Hayashi, K. Tada, K. Tsutsumida, K. Higeta, "External Memory BIST for System-in-Package", IEEE Intl. Test Conference, 2005, pp. 86 - 91

[8] J.C Chan, "An improved technique for circuit board interconnect test", IEEE Transactions on Instrumentation and Measurement Volume 41, Issue 5, Oct. 1992, pp. 692 - 698

[9] A. Cheng, C.C. Lim, A. Parashkevov, "A Software Test Program Generator for Verifying System-on-Chips", IEEE High-Level Design Validation and Test Workshop, 2005, pp. 79 – 86

[10] J. Perez, M. Sonza Reorda, M. Violante, "Early, Accurate Dependability Analysis of CAN-Based Networked Systems", IEEE Design & Test of Computers, Vol. 23, No.1, Jan/Feb. 2006, pp. 38-45

Test Programs	Detected					Not Detected	
	Data [#]	Data [%]	Timeout [#]	Timeout [%]	Total [%]	[#]	[%]
A	1,112	75.5	331	22.5	98.0	30	2.0
B	691	56.2	485	39.5	95.7	53	4.3
C	313	22.0	1,025	72.0	94.0	85	6.0
D	266	17.5	1,114	73.5	91.0	136	9.0

Tab 2: fault injection results.

1550-4093/07 $25.00 © 2007 IEEE

Circuit Profiling Mechanisms for High-Level ATPG

Jorge Campos and Hussain Al-Asaad
Department of Electrical and Computer Engineering
University of California, Davis, CA
E-mail: {jcampos, halasaad} @ece.ucdavis.*edu*

Abstract—Our Mutation-based Validation Paradigm (MVP) is a validation environment for high-level microprocessor implementations. To be able to efficiently generate test sequences, we need to enable MVP's ATPG to learn important details of the circuit under validation as a means to explore critical new circuit scenarios. In this paper, we present new profiling mechanisms that can exist either as a pre-processor that gathers circuit information prior to the circuit validation process, or as run-time entities that allow MVP to learn from its progressive experience.

I. INTRODUCTION

Our Mutation-based Validation Paradigm (MVP) [1][2][3][4] technology contains the fundamental techniques for analyzing high-level circuit implementations, and is unique in the way it exploits these techniques to validate circuit implementations. MVP's methods help deliver certainty into a circuit validation project in two ways: (*i*) It provides real-time observability into the validation effort through a concurrent mutant simulator that quantifies the circuit coverage (certainty level) at every simulation time-frame, and (*ii*) it employs deterministic circuit analysis techniques that, together with the observability provided by its concurrent mutant simulator, allow MVP's ATPG effort to consistently explore new corners in a circuit's architectural landscape. These contributions enable MVP to mitigate the risk of validation false-positives due to unexposed bugs, which is commonly encountered when random or pseudorandom ATPG fail to travel towards unexplored portions of the circuit under validation.

MVP is a circuit validation tool for high-level hardware descriptions, and its purpose is to provide expert deterministic validation methods to the average design engineer. MVP provides a complete and automated strategy for analyzing high-level hardware descriptions that only leaves the circuit design engineer to decide what portions of the circuit to validate, and not how to validate it. These circuit analysis abilities allow MVP to perform automated white-box circuit validation on high-level RTL descriptions while providing the simplicity of black-box validation to its users.

MVP does not require a priori information on the circuit under validation for it to be effective, but instead gathers this information real-time. Use of MVP's fundamental circuit analysis abilities alone cause it to be burdened by the analysis of irrelevant HDL code segments, and by the traversal of already-explored architectural states. We can significantly improve MVP's run-time performance by implanting mechanisms that enable it to *learn* important details of the circuit under validation as a means to explore critical new circuit scenarios. These mechanisms can exist as a pre-processor that gathers circuit

information prior to the circuit validation process, as well as run-time entities that allow MVP to learn from its experience.

MVP can handle complete implementations because it only uses high-level information, and only uses the hardware description language (HDL) information relevant to the set of constraints when identifying all relevant architectural states. In this paper, we define a circuit architectural state that satisfies the set of constraints under consideration as a *prospect state* (*pState*).

Generating input stimuli that satisfy a set of constraints requires the solver to identify all prospect states for each time frame, and eliminate the prospect states that can not be used to satisfy a test sequence. This is a problem for modern superscalar microprocessor implementations because of their inherently large state space. Therefore to be able to efficiently identify and analyze the architectural states (*prospect states*) that can possibly satisfy the set of constraints, we need to reduce the search space (via profiling) in the analysis process as early as possible.

The rest of this paper is organized as follows. Section II presents several pre-processor profiling techniques used by MVP and Section III presents the details of MVP's run-time profiling techniques. Section IV presents some preliminary experimental results and Section V concludes the paper.

II. PRE-PROCESSOR CIRCUIT PROFILING

The pre-processor to MVP's circuit validation process should be a light-weight task that provides MVP with valuable insight capable of directing its test pattern generation process towards a solution. It is because of this low-overhead demand that the pre-processor should not attempt to solve actual ATPG constraints, but rather solve early the sub-problems that provide MVP with the most valuable information. Instead of analyzing the implications that the circuit has onto each statement in the hardware description as is done in the real-time circuit analysis process (implications of the range of values for β and γ on the assignment to α as shown in Fig. 1a), we can implement the light-weight circuit profiler for the pre-processor by having it analyze the implications each statement has onto the overall circuit (assignment to α in Fig. 1b).

A. Assignment Statement Profiling

Our previous work discusses how an ATPG constraint is solved by exploring all relevant assignment statements that can satisfy its unresolved data implications [3]. Therefore whenever attempting to satisfy a constraint (especially when it is dependent on an enumeration data type), this solver process will likely be repeated for a great deal of assignment statements that cannot help satisfy the constraint. Much of this

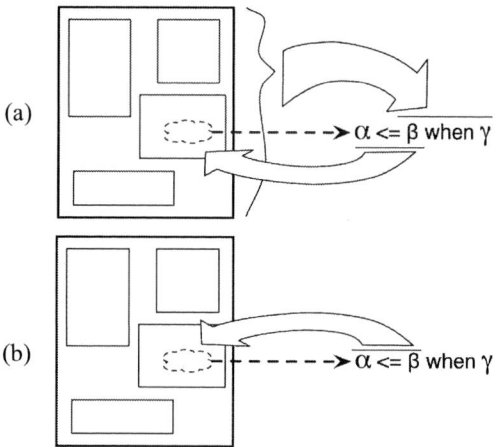

Fig. 1. (a) Run-time and (b) pre-processor circuit profiling.

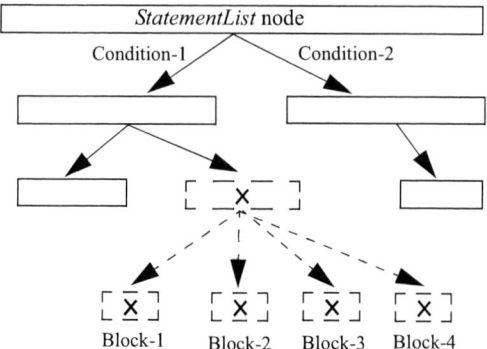

Fig. 2. Basic-block guard profiling.

dead-end work can be prevented by using MVP's *statementList* data structure [3], which holds a circuit's HDL information, to *index* each assignment statement with the identifier value implications that it has onto the hardware description. This *statementList* data structure is utilized by MVP to hold the HDL information of the circuit under validation.

The indexing process is easily implemented by using MVP's available resources. Satisfying a constraint requires MVP to first convert the assignment statement being considered into a data dependency graph (DDG) and to solve its implications. Solving this DDG provides every identifier within every conjunction with the explicit range in values that satisfies this assignment statement. Therefore to profile this specific assignment statement, the data implications it imposes onto the circuit are extracted directly from the identifiers within the solved DDG.

This indexing pre-processor is effective because many assignment statements in a hardware description simply transfer a constant value onto an identifier. This is particularly true for enumeration data types, as they are commonly used to explicitly control a finite state machine (FSM). This allows MVP to peek into each statement to expose most data contradictions before committing itself to solving that possible solution path.

B. Implicit Memory Element Profiling

For complete circuit analysis, MVP must explore all signals in the hardware description in search for implicit memory elements. It does this by negating the explicit guards to all assignment statements onto the signal being analyzed, and inserting them into a single conjunction (unified by Boolean AND operators). This process exploits MVP's efficient DDG solver, and a DDG that does not evaluate to false signifies an implicit memory element. This process therefore takes all implicit memory elements, and defines them explicitly by creating an entry for a corresponding memory-preserving assignment statement within the *statementList* data structure such that the guards to this entry denotes the memory-preserving condition.

C. Basic-Block Guard Profiling

In most cases where a data contradiction is encountered when solving a constraint, the contradiction arises from the union of the guards in the multiple prospect code paths. That is, the guards gathered from satisfying the constraint of a current unresolved identifier will more than likely conflict with the guard of a previously resolved identifier in the constraint. Experiencing an identifier value contradiction within the guard of a basic block is significantly more costly than experiencing a contradiction within the statement itself because the aggregated guards leading up to a basic block is larger in most cases than any of the assignment statements in that basic block, and this guard is repeatedly utilized by all statements within the basic block. Therefore the performance of this guard profiling pre-processor is slower than that of the assignment statement profiling pre-processor, but the runtime performance advantage it provides is equally as significant.

It is possible to take advantage of the *statementList* data structure once again to hold pre-solved identifier values from the guards to all basic blocks. These solved identifier values can be used to expose data contradictions between a prospect code path's guards and the identifiers within a constraint being solved, and would prevent the costly dead-end task of converting a set of guard statements into a corresponding solved DDG that would evaluate to 'false'. For those assignment statements that are reachable, this profiling effort can retain the solved DDG to optimize all later uses by any assignment statement within the basic block.

This process of indexing all leaf *statementList* nodes with the solved identifier values to its guards can be performed as a pre-processor or at run-time. Given that MVP already analyzes all identifiers to expose implicit memory elements, which requires it to evaluate the guards to all statements, it is natural to implement this basic block guard profiler as a part of the pre-processor. For any given *statementList* node, the set of guards are obtained by appending its guard to those of all its ancestor nodes [3]. We can take advantage of the fact that the guards are distributed throughout the *statementList* tree (Fig. 2) by gathering the list of solved identifier values at each node, and recursively providing a copy of this list to all its children so it may append onto it.

This recursive process to obtaining guard information provides us with two key advantages. The data-sharing nature of this recursive algorithm allows it, as a pre-processor that starts at the root *statementList* node, to reduce the amount of redundant work that would be performed if it were to be executed at run-time starting at a leaf node. The second advantage is that it can identify all unreachable basic blocks within the hardware description without having to analyze all basic blocks. As shown in Fig. 2, an identifier value contradiction exposed within the guard of an internal *statementList* node will automatically denote all its children statements as unreachable as well.

Unreachable code blocks commonly exist within CASE statements. The "when others" clause of a CASE statement is commonly utilized as a safe-guard that ensures an internal signal to the circuit is not created as an implicit memory element as a result of unhandled guard cases. However, when all possible cases have been handled explicitly, then the "when others" clause will create an unreachable block of code that will never be executed. Unfortunately for MVP, it attempts to satisfy a constraint by starting at all relevant assignment statements. MVP must therefore be aware of which HDL code blocks are irrelevant as a means of reducing the ATPG search space.

III. RUNTIME CIRCUIT PROFILING

MVP's run-time circuit validation process should be a complete task focused on exploring uncharted territory within the processor. In an ideal problem, it would be possible to travel throughout a hardware description's architectural state space without retracing one's steps. Unfortunately, 100% FSM coverage commonly requires a significant amount of redundant state exploration. Therefore as MVP gets further into its validation process, it will be forced to retrace more of the previously-explored state space in order to reach the target architectural state that defines the ATPG goal. Also, there are many architectural states that have a high occurrence frequency as they are a precursor to a wide range of other architectural states, thus retaining some of their pre-solved information can optimize MVP's performance in the long run. This section focuses on the run-time circuit profiling efforts that can allow MVP to breeze through the already-explored state space when attempting to satisfy a unique ATPG goal.

A. Finite State Machine Profiling

An FSM is usually described by $y = \delta(s, x)$, where a target state y can be reached from state s when the FSM's inputs are x [5]. When the target state y can be reached by multiple states $s_1 \ldots s_n$, we can use a weight scheme such that the state s with the lowest weight provides MVP with two advantages:

- When the pStates have never been explored (thus they are un-indexed), it will allow MVP to choose the state s with the least number of constraints that will need to be satisfied at the subsequent ATPG iteration. If the reset state is among the set, it will be characterized by the lack of constraints that need a subsequent ATPG iteration, therefore resulting in a weight of zero.
- When any of the pStates has been previously explored, its weight will be lower than all unexplored pStates, and will provide MVP with guidance towards the reset state as all subsequent pStates will continue to have lower weights.

To implement this weight-assigning process, we simply need to implement our ATPG solver as shown in Fig. 3. Line 5 selects the optimal candidate for the next ATPG iteration by selecting the pState with the lowest weight. If the selected pState has a weight of zero, the previous recursive call to the *multiFSM_solve* algorithm has its length value l updated to

```
multiFSM_solve(testSequence TS, pStateSet Y, int l)
1.   If (l = 0) return FAIL
2.   If (Y = reset state)
3.      return SUCCESS
4.   P ← solve(Y)            //Generates set of solutions
5.   For each t ∈ P, s.t. t has the lowest weight in P
6.      If (weight(t) = 0)
7.         l ← 0
8.         return SUCCESS
9.      S ← get_previous_timeFrame(t)
10.     If (multiFSM_solve(TS, S, l – 1) = SUCCESS)
11.        assign_weight(S, l)
12.        l ← l + 1
13.        TS ← TS + getPrimaryInputs(S)
14.        return SUCCESS
15.  return FAIL
```

Fig. 3. MVP's optimized ATPG algorithm.

zero on line 7, and it is returned SUCCESS signifying that the reset state has been reached on line 8. The previous recursive call to the *multiFSM_solve* function will then be in charge of updating the weight values on line 11, incrementing the weight for its previous recursive call on line 12, and then commencing as usual.

The goal of exploring an FSM is to generate a test sequence that maps the hardware description's architectural state from its reset state onto any architectural state that satisfies the given set of constraints. This process begins at the target architectural state, and continues to traverse the circuit backwards in time until the reset state is reached. To optimize this ATPG effort, we need to enable MVP to intelligently navigate through a circuit's FSM.

A hardware description is characterized by the inter-dependent FSMs from all of its internal registers, thus developing a macroscopic understanding on the overall FSM will require us to understand all possible state combinations (the cross product) from all these smaller inter-dependent FSMs. We can therefore simplify the FSM profiling task by placing our focus at the individual FSMs for each register as they make up the building blocks for the overall FSM.

Our objective in performing FSM profiling on the overall circuit is to achieve the profiling tasks on the individual FSMs, and employ a mechanism that translates this low-level FSM profiling information into a circuit-wide FSM profiler. The concept is simple enough, but the implementation is tricky because MVP does not manage these FSMs explicitly. It would be possible to provide MVP with the mechanisms that allow it to build and analyze these interacting FSMs explicitly, but that would only require it to perform another level of computations that should not be necessary. MVP's strength is in its ability to analyze the circuit under validation by focusing on the source code, and it is possible to exploit MVP's source code database of the circuit under validation to achieve similar profiling results.

Let us take a moment to translate these FSM profiling goals into MVP's language. The low-level FSM profiling is meant to account for the many inter-dependent FSMs, and so it must therefore analyze the FSM associated with each identifier that represents an internal register. MVP currently uses a construct entitled as an *identifierSet*, whose purpose is to keep track of every HDL location that each identifier is assigned a value

1550-4093/07 $25.00 © 2007 IEEE

onto. Initially, the objective of this construct was to optimize the algorithm that generates all possible pStates from a given identifier constraint by having the sources to all possible solutions be readily available in one data structure. Therefore, we can also use all entries corresponding to a constraint's identifier to provide us with the FSM profiling information we need. We can exploit the fact that MVP accesses this *identifierSet* data structure each time it attempts to use a code path as a solution by also having MVP leave behind real-time low-level circuit profiling information whenever it successfully utilizes this data source to satisfy a constraint.

The aforementioned global FSM profiling effort is meant to interpret the low-level FSM profiling information and identify the shortest FSM path that can reach the circuit's reset state. We know the low-level profiling effort should be performed when MVP attempts to use a line of HDL source code for satisfying a circuit constraint, therefore we should take a step back and identify which MVP construct is analyzing these lines of code and could stand to benefit from the low-level profiling efforts. Looking at Fig. 3, we can see that the resulting test sequence is generated by instantiating pStates as the mechanisms that carry the potential solutions as they are being generated, and thus the pState construct should be used to manage the global FSM profiling effort.

The low-level FSM profiling effort is focused on depositing information onto each statement in the hardware description to record its scope and the success it provides. Conversely, the global FSM profiling effort is focused on unifying the information gathered from all statement sources that represent a given solution as a means to avoid costly or irrelevant scenarios. The remainder of this section revolves around these concepts.

FSM Weight Indexing. MVP's ATPG algorithm is able to independently find the reset state through exploration of an FSM, but this alone requires much backtracking. We can therefore exploit its ability to find and detect the reset state by appending the explored states in each FSM (the explored assignment statements for the identifier behind the FSM's register) with a weight value equal to its distance from the reset state. If MVP is instructed to generate a test sequence with a length of at most l, then we can assign each state an initial weight $\gg l$.

Unfortunately, the task of assigning weight values to a processor's architectural states is not so straightforward. This is because each pState is influenced by multiple implicit FSMs, and is pieced together by several concurrent assignment statements that successfully satisfy all simultaneous constraints. MVP, therefore, is not assigning weight values to explicit architectural states, but rather is assigning weights to the assignment statements that were used to piece them together.

MVP can perform its run-time weight-assigning process following every ATPG iteration to update each assignment statement's resulting distance to its FSM's reset state. Any given assignment statement may impact several distinct architectural states, and thus its weight value may have multiple sources. For the sake of allowing MVP to move towards an optimal solution while keeping the ATPG implementation simple, we will allow each assignment statement to store the lowest weight value it is assigned. Using a given assignment statement's lowest assigned weight value, say w, is reasonable because that statement has the potential of providing an

instruction sequence of size w again in the future. Therefore giving preference to this assignment statement over other alternate assignment statements of higher weight when solving a constraint allows MVP to choose the ATPG path with the highest probability of producing the shortest path to the reset state.

Prospect State Weight Estimation. MVP's ATPG algorithm analyzes multiple pStates at every time frame, from which it must choose one to attempt and reach the reset state. Therefore providing MVP with a weighing scheme for its pStates can help it easily identify the most effective solution path. The motivation for extracting a weight value from a pState is two-fold, as mentioned at the start of this section. In choosing the ideal pState, MVP must first favor those solutions to which a path to the reset state has already been identified; otherwise it must favor the pStates with the least number of constraints to justify. These two objectives must be handled inherently by a single weighing scheme.

However, finding a balance between these two objectives is not trivial because the first requires that the pState have been solved in order to extract an accurate weight from the utilized HDL statement sources, and the second requires the pState to *not* have been solved. Using our FSM weight indexing scheme where we index each RTL assignment statement with its known distance to the reset state, we can attain a weight value to a solved pState because it will then have assignment statements associated to it that were used to satisfy its constraints. Thus for the first case, if a pState has not been solved, then it will not have these HDL statement sources that are necessary to estimate its distance to the reset state. Conversely for the second case, the number of constraints to resolve in a pState obviously can only be evaluated before these constraints are resolved.

In identifying a pState's weight, MVP must use a unifying scheme that satisfies both of the preceding objectives. MVP will first have to solve the pState, and then adapt its weight-assigning scheme to handle the second case which favors the pState with the least number of ATPG constraints. It can do this adaptation by counting the number of constraints that will propagate onto the following ATPG iteration. And to estimate the weight that gives preference to those previously-solved pStates closest to the reset state, we can multiply this number of constraints that need to be resolved in the next ATPG iteration by the average weight of the assignment statements associated to the solved constraints. A pState whose constraints were solved in a previous ATPG problem will have assignment statements associated to it whose weight is lower than the maximum weight, and thus its average weight will naturally be lower than the maximum weight.

Modified ATPG Algorithm. MVP's pState-weighing scheme requires us to modify MVP's ATPG algorithm as depicted in Fig. 4. The *get_previous_timeFrame()* algorithm extracts, from a pState y, all the pStates s that can transition into it. It requires y to have been solved (have all its constraints satisfied), and it returns a set of pStates s that have not been solved. Thus, the objective of this modification is to ensure that MVP's ATPG algorithm calls the weight estimation procedure on solved pStates only, as well as perform weight indexing using these solved pStates.

The most significant change that allows us to satisfy these objectives is that the algorithm now expects the alternate

Precondition: pStates in *Y* have been pre-solved

```
multiFSM_solve(testSequence X, pStateSet VS, pStateSet Y, int l)
1.   If (l = 0) return FAIL
2.   For each p ∈ Y, s.t. p has the lowest weight in Y
3.       P ← get_previous_timeFrame(p)
4.       For each p' ∈ P
5.           If (p' is masked by some state in VS)
6.               delete p'
7.           else
8.               VS ← VS + p'      // Store copy of p' into visited set VS
9.               If (weight(p') = 0)      // Reset state has been found
10.                  l ← 0
11.                  Y ← {p}      // Remove unexplored pStates in Y
12.                  X ← getPrimaryInputs(p')
13.                  return SUCCESS
14.              S ← solve(p')      // Attain incoming pStates S from
                                    // target pState Y
15.              If (multiFSM_solve(TS, VS, S, l − 1) = SUCCESS)
16.                  weight(p) ← l
17.                  Y ← {p}      // Remove unexplored pStates in Y
18.                  l ← l+1
19.                  X ← X + getPrimaryInputs(S)
20.                  return SUCCESS
21.  return FAIL      // No ATPG goals (Y is empty) or
                      // no solution exists
```

Fig. 4. MVP's modified ATPG algorithm.

ATPG objectives *Y* to be a previously solved set of pStates. Having *Y* be a solved set of pStates allows MVP to immediately use its weight estimation methods for identifying the ATPG goal in *Y* that is estimated to be closest to the circuit's reset state. Afterwards, this modification converts the chosen path in *Y* into the alternate sets of constraints *P* that define the preceding architectural states. If the pState set in *P* contains the reset state, then the ATPG iteration is complete. Otherwise the set in *P* is solved to define the set of previous time frames *S* that can transition into *Y*, and to define the inputs that allow this transition to take place. The preceding pStates in *S* are themselves justified towards the reset state by invoking a recursive call to the ATPG algorithm.

B. Explored State-Space Tracking

The ATPG algorithm in Fig. 4 will commonly receive, from line 3, pStates that have been traversed by a previous recursive call within the same ATPG iteration. When this happens, those pStates should be ignored because re-analyzing them will not help the ATPG algorithm get any closer towards a solution. Ignoring the visited pStates is both an up-stream and downstream process. Preventing the ATPG algorithm from revisiting a pState that is visited earlier in the same test sequence will prevent the ATPG algorithm from analyzing FSM loops. Furthermore, preventing the ATPG algorithm from revisiting a pState that was visited by a previous test sequence branch that failed to generate a result will prevent the ATPG algorithm from analyzing unsuccessful paths more than once.

This explored state-space tracking effort is implemented on lines 5, 6, and 8 of Fig. 4. Line 5 checks if the current pState *t* to be analyzed has been previously visited by that same ATPG iteration. If it has been previously visited, then line 6 deletes it and allows the subsequent iteration of the FOR loop on line 4 to analyze the next pState in the solution set *P*. If it has not been previously visited, then line 8 allows the ATPG algorithm to store *p'* into the visited set *VS* and proceed as usual.

We can identify if a pState *p'* has been previously visited by identifying if *p'* is masked by the set of visited pStates in *VS*. A pState is defined by a set of internal and input identifiers, and their corresponding range in values. For the purpose of obtaining a clear perspective on when one pState masks another, let us realize that an identifier missing from a pState signifies that the corresponding identifier has a complete range in values. In terms of identifiers, an identifier with a range in values *v* is masked by a corresponding identifier instantiation with a range in values *v'* if and only if (IFF) the range in values for *v* are encapsulated by the range in values for *v'* ($v \in v'$). We can therefore identify if a pState *t'* is masked by a pState *t* IFF the set of identifiers referenced by pState *t* is a subset of the identifiers referenced by *t'*, and IFF the range in values of the identifiers in *t* encapsulate the range in values of the corresponding identifiers in *t'*.

IV. EXPERIMENTAL RESULTS

Results on MVP's effectiveness have been generated by following the key steps in a validation paradigm: coverage metric definition, error modeling, circuit simulation, and ATPG. All experiments have been performed on a Dual 2.5GHz G5 workstation under OS X Tiger using gcc 4.0. MVP has been implemented as a library using GNU's autotools (*autoconf, automake, libtool*) in 20K physical lines of C++ code.

The Motorola 68K implementation [6] analyzed for this paper is microarchitectural by nature. The collection of mutation control errors (MCEs) [1][2] are generated to bind all possible combinations between the explicit *state* signal to all other control signals. The set of control signals also includes the *next_state* signal, which allows us to stimulate the data paths as well as the FSM transitions. Having a combination of values between a control signal (imagine an input to a multiplexer) and the explicit state as ATPG constraints allows us to stimulate every combination of the control signal's set of resulting data paths at every explicit microarchitectural state. Furthermore, having a combination of values between the *state* signal and the *next_state* signal allows us to stimulate every transition in the microarchitectural FSM.

The Motorola 68K explicit FSM allows us to easily understand MVP's effectiveness in deciphering a circuit's state machine by observing the shortcuts it exploits when satisfying a set of constraints. Each of MVP's ATPG iterations generates a test sequence that maps any target architectural state to the reset state. However if the reset state has not been identified, the search process is blind. After even one successful ATPG iteration, MVP's effectiveness in reaching the reset state is highly optimized by its run-time circuit profiling methods. We can quantify MVP's effectiveness in learning a circuit's FSM by counting the amount of backtracking experienced by each ATPG iteration. Fig. 5 quantifies how suddenly the backtracking is reduced by MVP's run-time circuit profiling methods after every successful ATPG iteration.

MVP's true effectiveness is due to its ability to continuously traverse unexplored portions of a circuit's architectural state-space. The circuit design industry currently employs random and pseudo-random ATPG when exploring large circuit imple-

1550-4093/07 $25.00 © 2007 IEEE

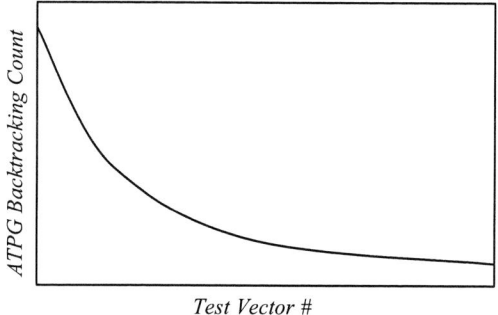

Fig. 5. MVP's FSM-learning effectiveness.

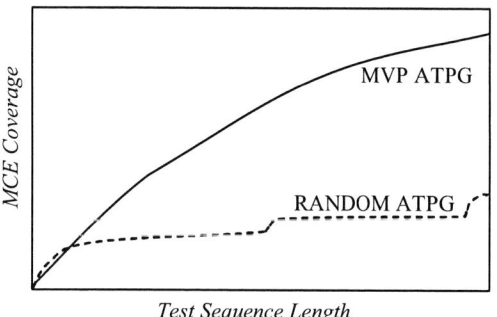

Fig. 6. MVP's ATPG effectiveness.

mentations. This provides them with a high rate of simulation iterations, but they gamble on their random approach to eventually reach a high coverage of their circuit's state-space. By applying a given set of mutants onto MVP's concurrent mutant simulator, we can directly compare the effectiveness in MVP's approach to the random methods commonly used to expose circuit design errors. Fig. 6 presents the effectiveness provided by MVP in comparison to random ATPG.

If we look closely at the start of the simulation in Fig. 6, we can see that random ATPG is initially more effective than MVP's deterministic ATPG. The reason for this is because every test sequence generated by MVP begins at the circuit's reset state. As a result, the initial test vectors in each of MVP's ATPG traverse already-explored architectural states. This is acceptable because using the reset state as the common test sequence starting point provides us with two advantages: (*i*) The ATPG unit is able to perform real-time profiling such that each HDL line of code can hold its weight with respect to its known shortest distance to the reset state, and (*ii*) any given test sequence that exposes an actual circuit design error is self-sustained and can be utilized independently, as it begins with the circuit's reset state and ends when the circuit design error is exposed.

By looking closely towards the end of simulation in Fig. 6, we can see that MVP's deterministic ATPG is significantly more consistent and effective than a random or pseudo-random ATPG approach. The simulation results from the random ATPG soon levels off, and has eventual bursts of effectiveness whenever the random test vectors happen to reach an unexplored portion of the architectural state-space. These sudden burst of productivity cannot be predicted, and are commonly a

source of false-positives in circuit validation. MVP's deterministic ATPG, however, exploits MVP's simulation statistics to determine the ATPG goals that can allow it to continuously reach unexplored portions of the state-space. Therefore MVP, unlike random and pseudo-random ATPG, has *consistent* bursts of productivity due to MVP's closed-loop strategy between circuit simulation and automated test pattern generation.

V. Conclusions

We have presented circuit profiling mechanisms that allow our mutation-based validation system to learn as it generates test sequences. These mechanisms are either a pre-processor that gathers circuit information prior to the validation process or a run-time entity that progressively gathers circuit information during the validation process. Our preliminary experiments show that MVP's effectiveness in reaching the reset state is highly optimized by its circuit profiling methods. Moreover, the experiments show that the backtracking in MVP's ATPG is reduced by using run-time circuit profiling methods after every successful ATPG iteration.

Acknowledgment

This material is based upon work supported by the National Science Foundation under Grant No. 0092867.

References

[1] J. Campos and H. Al-Asaad, "Concurrent design error simulation for high-level microprocessor implementations", *Proc. AUTOTESTCON*, 2004, pp. 382-388.

[2] J. Campos and H. Al-Asaad, "Mutation-based validation of high-level microprocessor implementations", *Proc. International High-Level Design Validation and Test Workshop*, 2004, pp. 81-86.

[3] J. Campos and H. Al-Asaad, "MVP: A mutation-based validation paradigm", *Proc. International High-Level Design Validation and Test Workshop*, 2005, pp. 27-34.

[4] J. Campos and H. Al-Asaad, "Search-space optimizations for high-level ATPG", *Proc. International Microprocessor Test and Verification Workshop*, 2005, pp. 84-89.

[5] F. Corno et al., "SymFony: A hybrid topological-symbolic ATPG exploiting RT-level information", *IEEE Transactions on Computer-Aided Design*, Vol. 18, pp.191-202, February 1999.

[6] http://www.opencores.org/projects.cgi/web/system68/overview.

[7] I. Ghosh and M. Fujita, "Automatic test pattern generation for functional register-transfer level circuits using assignment decision diagrams", *IEEE Transactions on Computer-Aided Design*, Vol. 20, pp. 402-415, March 2001.

[8] J. Shen and J.A. Abraham, "An RTL abstraction technique for processor microarchitecture validation and test generation", *Journal of Electronic Testing: Theory and Applications*, Vol. 16, pp. 67-81, February-April 2000.

[9] L.-C. Wang and M.S. Abadir, "On efficiently producing quality tests for custom circuits in PowerPC[TM] microprocessors", *Journal of Electronic Testing: Theory and Applications*, Vol. 16, pp. 121-130, February-April, 2000.

[10] F. Corno et al., "Automatic test program generation from RT-level microprocessor descriptions", *Proc. International Symposium on Quality Electronic Design*, 2002, pp. 120-125.

[11] M. N. Velev, "Collection of high-level microprocessor bugs from formal verification of pipelined and superscalar designs", *Proc. International Test Conference*, 2003, pp. 138-147.

[12] C.-C. Yen, J.-Y. Jou, and K.-C. Chen, "A divide-and-conquer-based algorithm for automatic simulation vector generation", *IEEE Design and Test of Computers*, Vol. 21, pp. 111-120, March-April 2004.

[13] D. Moundanos, J. A. Abraham, and Y. V. Hoskote, "Abstraction techniques for validation coverage analysis and test generation", *IEEE Transactions on Computers*, Vol. 47, pp.2-14, January 1998.

1550-4093/07 $25.00 © 2007 IEEE

Functional Test Selection for High Volume Manufacturing

Vijay Gangaram, Deepa Bhan, James K Caldwell, Intel Corporation, Folsom, CA

Abstract— Validation and legacy test suites are often reused for achieving at speed coverage required for testing high frequency semiconductor chips. Porting validation tests to high volume manufacturing (HVM) flows involves extensive manual effort but is required to ensure high quality chips. Functional test selection is the problem of choosing a subset of tests from a large pool of existing tests to maximize the fault coverage while minimizing the test data volume, fault grading time and porting effort. We formulate a framework for test selection that allows various coverage metrics to be used for evaluation. A novel dynamic untestability analysis method is proposed to identify faults that can not be detected by a given test sequence. Conversely this method can be used to compute tight upper bound coverage and hence as a metric for functional test evaluation. Test selection using this new metric gives significant additional fault coverage than toggle based test selection.

Index Terms—Functional Test Sequences, Design Validation, Test Sequence Compaction, Fault Simulation Acceleration, Untestable Fault Identification

I. INTRODUCTION

Semiconductor devices are becoming increasingly complex in terms of transistor count, frequency and integration. Multi-core designs and System-On-Chip are emerging trends in the industry. These coupled with aggressive design methods to optimize for power and frequency further pose significant challenges for testing. These challenges include testing for manufacturing defects as well as speed classification (binning) of devices such as microprocessors. Also studies have shown additional fallout for functional tests even with high scan (structural) coverage [1,2]. Due to these reasons, functional tests running on structural and at-speed ATE continue to play critical role in ramping high performance CPUs to market.

Design validation and verification is the process of ensuring correctness of a design at different levels of abstraction during the design process. While formal verification techniques such as logic equivalence checking and property verification are becoming common during later stages of the design cycle, dynamic (simulation based) validation continues to be the primary vehicle in design validation. In addition to applying automated test generation techniques [3,4,5], designers spend significant effort to develop functional test sequences that exercise the design and build confidence that the design matches specification. These functional test sequences can be quite large, several thousands for a typical microprocessor and

long, each ranging from a few thousands to millions of cycles. Several coverage metrics have been proposed in the literature to evaluate functional test sequences [6,7,8], but only a few [7, 8] touch upon observability but involve significant complexity.

Functional tests developed during design validation stage as well as HVM tests from previous generation of the microprocessor can be reused for manufacturing testing. Due to test time and tester memory limitations, only those tests that provide good manufacturing defect screening value are added to production test tape. Fault simulation for specific fault models such as stuck-at, transition and bridge are often used to grade the functional test sequences. Selected functional test sequences may need to be modified to ensure they are compliant to application on the tester. The process of selecting tests from a large pool of functional test sequences is called functional test selection.

Current approaches to functional test selection for manufacturing testing are ad-hoc and often use rough coverage metrics such as toggle coverage and/or prior knowledge. These functional test sequences are very long with total test suite exceeding billions of simulation cycles. Exact methods such as fault simulation and even statistical fault simulation of the entire test suite are not practical due to computational costs. We propose a fast and accurate coverage metric based on dynamic untestability analysis and show how it can be used in test selection to achieve more fault coverage than what is possible with toggle metric.

The rest of the paper is organized as follows. We first describe our test selection algorithm that is general to accommodate various coverage metrics. Then we explain the toggle coverage metric and its limitations. We then describe the new dynamic untestability analysis method and show how it compares with toggle based test selection against the ideal method of fault simulation based test selection.

II. FUNCTIONAL TEST SELECTION FRAMEWORK

Given a set of functional test sequences $\{T1...TN\}$ and a design with a set of faults F, a test selection scheme should select a minimal subset of tests in an ordered fashion such that all faults in F are covered.

Definition: A fault Fi is considered *covered* by a test Tj if the fault is detected by the test as measured by the coverage metric being employed.

The definition for a fault varies based on the coverage

metric used. In case of toggle coverage metric, the fault set F would be defined as a set of all RTL variables and condition branches and a fault is considered detected by a test if it is toggled by the test. Similarly, in case of stuck fault coverage metric, the fault set is the universe of stuck faults in the design.

One can see that test selection problem can be translated to a set covering problem. We use a simple greedy heuristic (Figure 1) to select tests until cumulative coverage is maximized.

Figure1: Test Selection Algorithm

1. Estimate Coverage Metrics (faults covered) by each test.
2. Compute unique coverage of each test
3. While faults left to be covered
   a) select test with the most unique coverage
   b) if no test or multiple tests, select one with the most incremental coverage
   c) if there is a tie select one with the least cycle count
   d) mark faults detected by the test as covered

The proposed test selection algorithm is general and can be used with any coverage metric and/or fault model. We explain this further with toggle coverage as the metric.

A. Toggle Coverage Based Test Selection:

Validation test sequences are typically targeted tests and activate only portions of the design that are being targeted. We refer to percent of RTL signals and condition banches that are toggled by a test sequence as toggle coverage for that test. Correlation between toggle coverage and stuck-fault coverage varies based on HDL abstraction, but as shown in figure 2, there is reasonable correlation for validation test sequences. This is particularly true in designs with sufficient observability (test points) on the internal signals. Hence one can use toggle coverage metric as a proxy for fault coverage and perform test selection.

We can expect to generate toggle data for each test as part of design validation. It contains the number of times each node toggles to 0 and to 1. This data can be gathered from any standard HDL simulator with a small performance overhead. Fault set here is a list of all nodes in the design. Toggle coverage is defined as the set of nodes that toggled to 0 or to 1. The goal of toggle based test selection is to make sure that every node is covered by at least one test. A unique detect is a node (fault) that is toggled (detected) by only one test.

Figure 2: Toggle and Stuck Fault Correlation for functional tests

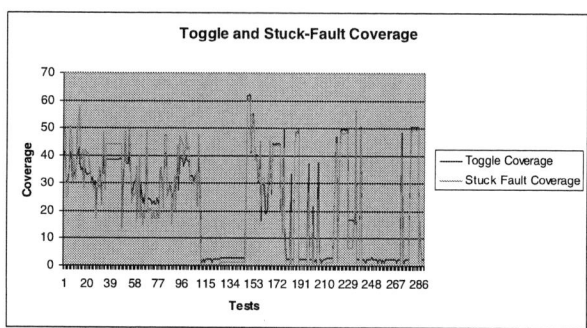

While toggle coverage is easy to compute, the metric is not effective in terms of inclusiveness or precision when they are measured in terms of fault coverage.

Observation 1: Toggle metric can be pessimistic. It is possible for a node to not toggle (say constant 0) throughout a test sequence, yet the stuck (at 1) fault is excited and possibly detected. Toggle coverage based test selection could incorrectly skip such a test since it doesn't toggle the node of interest.

Observation 2: Toggle metric is more optimistic than stuck fault coverage since it does not consider observability. A test that toggles a node will excite the fault on that node, but detection might be blocked. Toggle metric fails to provide such insight.

In general, toggle coverage tends to be optimistic and hence the test selection saturates too early to maximize fault coverage. Ideal test selection for HVM would use exact fault simulation as the metric on all possible faults. While it can achieve all of the coverage that is possible with the test suite, it is simply not practical due to excessive fault simulation run times. Statistical fault simulation suffers from inaccuracy unless fault sampling and grading is repeated after each test selection which in turn means exorbitant run times.

If we have to use a coverage estimate, it is preferable to use a coverage metric with no pessimism for selection and a different but more pessimistic metric for taking credit (step 3.d in Figure 1). This will allow the test selection to not skip valid tests from consideration there by achieving high fault coverage at the expense of suboptimal number of tests to be selected. We provide such a scheme next.

III. DYNAMIC UNTESTABILITY ANALYSIS

We will introduce a few definitions before describing the details of the method.

Definition: A signal is considered *0(1)-uncontrolled* if it was never set to 0 (1) during the simulation of a functional test sequence.

Definition: A fault is considered *pattern-unexcitable* if the fault is not be activated by the test. For example, stuck-at-0 fault on a node that is 1-uncontrolled is pattern-unexcitable.

Definition: A fault is *pattern-unobservable* if its propagation is always blocked during the test.

Definition: A fault is *pattern-untestable* if the fault is proven untestable under the pattern specific conditions. This set includes both pattern-unexcitable and pattern-unobservable faults.

As previously mentioned, functional test sequences are typically targeted tests and toggle only a portion of the design. This means that a large number of nodes are not controlled to certain values during the *entire* test sequence. Given the initial state of the circuit in RTL simulation and toggle data, one can deduce the set of 0(1)-uncontrolled nodes by a given test.

Given this uncontrollability information, we perform untestability analysis to identify faults that are not detectable by the test. Conversely, faults that are not untestable under the given uncontrolled conditions are potentially detectable by the test. We call this *Coverage by UNTestability* analysis method or COUNT metric. This is similar to fault undetectability determination from logic simulation by [], but extends the concept to entire test sequence as opposed to per vector analysis on combinational circuits.

Definition: COUNT metric computes the fault coverage of a test as the percent of faults that are not untestable given the uncontrolled conditions that are specific to a test. COUNT metric provides an upperbound on coverage and has no pessimism.

$$DetectableFaults = TotalFaults - PatternUntestableFaults$$

$$FaultCoverage = \frac{(100 * DetectableFaults)}{TotalFaults}$$

This is better explained with an example. Consider the circuit in Figure 3 where two operands are multiplied by MULT and results are stored in a register file (R1 or R2) and read out at a later time. Let us say that a functional test sequence (T1) performs multiplication, writes to register R2 only but does not read from it. In other words, Read Enable and Write Address signals are 1 and 0-uncontrolled respectively and block all faults in the multiplier regardless of the toggle coverage with in the multiplier.

Figure 3: Example for Dynamic Untestability

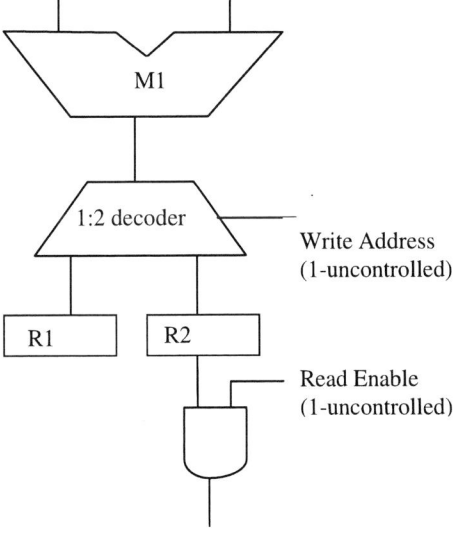

COUNT can utilize any algorithm that is effective in finding untestable faults. We use a set of algorithms that are fast and effective. Figure 4 captures the algorithmic details of COUNT.

Figure 4: COUNT Algorithm

1. Extract Uncontrollable signal values from Logic Simulation.
   a. Add unexcitable faults to pattern-untestable faults
2. Unobservable Gate Identification
   a. Add unobservable faults to pattern-untestable faults
3. Constrained combinational ATPG
   a. Add redundant faults to pattern-untestable faults
4. Report Fault Coverage

A. Extracting Uncontrolled signal values

We extract uncontrolled signal values from HDL simulation of the functional test. As a result, signals that are internal to HDL expressions are not visible. We used forbidden value simulation [9] to deduce uncontrolled signal values on internal nodes. For example, an AND gate output will be deduced as 1-uncontrolled if one of its inputs is 1-uncontrolled.

B. Observability Analysis:

Uncontrolled signal values can block the observability of other gates. For example, Read Enable signal in figure 3 blocks observability of all upstream faults. We perform a depth-first-search starting from observable outputs and recursively mark input gates that are observable. This technique is similar to [10,11] but applied once for the entire test sequence as opposed to every vector. Also we perform propagation path check on fanout stems to check if they are observable. Gates that are not marked as observable at the end of DFS are treated as unobservable and faults on them are added to pattern-untestable fault set.

C. Constrained Combinational ATPG

We perform combinational ATPG where the search space is limited by uncontrolled signal values in the good machine. Faults that are proven redundant here are added to pattern-untestable fault set. There is a possibility that inputs to the combinational logic (sequential elements) are on fault-path. We determine this sequential fault path information with the help of a state reachability graph where there is an edge between two states A and if there is a path (combinational or sequential) from A to B. This can be built using an algorithm that has worse case complexity of O(M*N) where M is the average combinational cone size and N is the number of sequential elements. Details of the algorithm are skipped for space consideration.

D. COUNT-S (Safe)

Since these techniques take multiple path sensitization and fault-path into account, faults that are considered pattern-untestable are indeed not detected by fault simulation. Therefore COUNT metric is safe and provides an upper-bound coverage. Another side benefit is speed up of fault simulation by removing these pattern-untestable faults[11]. This benefit is quantified in the results section.

E. COUNT-P (Pessimistic)

The disadvantage of above COUNT metric is that it provides upper bound coverage and therefore the optimism can cause the test selection to saturate prematurely. It's observed that multiple-path sensitization is rarely required for fault detection. This approximation is used in modifying COUNT algorithm to be less optimistic. Step 3 of COUNT algorithm (Figure 4) is modified in the following manner:

1. Find pattern specific unobservable states. A state is considered unobservable if both its stuck-at-0 and stuck-at-1 faults can not be detected. This is determined using above untestable fault identification algorithms.
2. Build a state graph where there is an edge from state A to state B if there is a combinational path that is not through any unobservable gates.
3. Starting at primary outputs, perform depth-first-search via observable gates and states and mark observability depth. It is the shortest distance from the state node to a primary output.
4. Mark states with observability depth greater than a threshold (we used a depth of 4) as unobservable.
5. While applying constrained ATPG on target faults, consider only primary outputs and those state elements that are not marked as unobservable.

During test selection, we use COUNT-P Metric to take credit for a test that is selected using COUNT-S metric.

IV. RESULTS

We performed our experiments on 4 design portions of a microprocessor. The design characteristics are captured in figure 5. Although there are thousands of functional test sequences available for each circuit, we used only a subset of tests that target each one so as to reduce the fault simulation run times.

Figure 5: Design and Test Characteristics

Design	#of Gates	#of Tests	Total Stuck Fault Coverage	Avg Test Sequence Length
Ckt1	238998	110	91.08	353318
Ckt2	145986	181	73.11	68319
Ckt3	127874	221	93.79	390224
Ckt4	43372	314	77.12	68517

Toggle data is derived from HDL simulation of each test sequence. Uncontrolled signal values are derived from untoggled signals using the initial state of the circuit. Gate level models synthesized from HDL are used for computing various coverage metrics (toggle, COUNT and stuck-fault simulation). Same fault sites are used in all three metrics to ease coverage metric evaluation. Toggle data for internal (synthesized) signals is derived based on toggle data for HDL signals. This extends HDL toggle coverage to be gate toggle coverage and gives the effect of statement and condition coverage in the HDL. We have observed that results are worse for toggle based test selection if only HDL nodes are used.

Run time requirements are shown in figure 6. COUNT adds a very small run time overhead on top of toggle simulation, but it comes with a good side benefit that faults marked as undetected by COUNT do not need to be fault simulated. This percentage fault pruning is significant and varies 32% to 72%. Note that all these are pattern specific untestable faults and are not real untestable faults.

Figure 6: Run Time Details

Design	Average Run Time (sec)			Faults Pruned by
	FSIM	Toggle	COUNT	COUNT (%)
Ckt1	17433	995	98	32.52
Ckt2	41880	2755	83	72.67
Ckt3	37095	2314	112	52.96
Ckt4	1616	15	13	39.20

Finally, the test selection results using toggle and COUNT metrics are captured in figure 7. COUNT metric achieves most of the stuck fault coverage possible. The loss in fault coverage (difference between column 4 of figure 5 and stuck fault coverage of the tests selected) is shown in parenthesis. While toggle coverage provides decent fault coverage, the metric saturates early losing coverage on the hard to detect faults. Fault coverage results for same number of COUNT tests as toggle are shown in column COUNT*. Surprisingly good results for toggle based test selection in spite of lack of observability are due to 1) sufficient observability (test points) on internal signals and 2) targeted nature of functional tests in the test pool.

Figure 7: Test Selection Results

Design	# of Tests (Toggle / COUNT)	Stuck F.C of tests selected using		
		Toggle	COUNT	COUNT*
Ckt1	51/58	90.35(-0.73)	90.50(-0.58)	90.38
Ckt2	53/118	71.91(-1.20)	73.08 (0.03)	72.03
Ckt3	42/62	91.34(-2.45)	93.77 (0.02)	93.30
Ckt4	23/71	72.88(-4.24)	75.85(-1.27)	72.80

V. CONCLUSIONS AND FUTURE WORK

We introduced the problem of functional test selection for HVM testing along with a general test selection framework. We proposed a novel dynamic untestable fault analysis concept and showed its benefit over toggle via test selection. Results validate its merit for maximizing fault coverage in test selection as well as a preprocessing step to accelerate fault simulation. There is further work in progress to improve the accuracy of dynamic untestability analysis based on additional information from HDL simulation. The proposed method can also be used as a static compaction method for test sequences. Test selection and dynamic untestability analysis presented in this paper can be extended for other fault models. Although the focus in this paper has been on manufacturing testing, the approach can be used in design validation to reuse validation tests for a previous processor as well as to tests that should be rerun on a changed design.

REFERENCES

[1] Eichelberger, E.B., T.W. Williams, "A Logic Design Structure for LSI Testability," Proc. Of 14th Design Automation Conference 1977, pp. 462-468.

[2] Maxwell, P.; Hartanto, I.; Bentz, L., "Comparing functional and structural test", Proceedings of International Test Conference, 2000,. Pp 400-407

[3] Nigh, P.; Needham, W.; Butler, K,; Maxwell, P.; Aitken, R., "An experimental study comparing the relative effectiveness of functional, scan, IDDQ and delay-fault testing", VLSI Test Symposium 1997, pp 459-464

[4] Parvathala, P.; Maneparambil, K.; Lindsay, W.; "FRITS - a microprocessor functional BIST method," International Test Conference, 2002. Proceedings 2002, pp 590 - 598

[5] Fummi, F.; Pravadelli, G.; Toto, F, "Coverage of formal properties based on a high-level fault model and functional ATPG", European Test Symposium, 2005. pp 162 - 167

[6] Santos, M.B.; Goncalves, F.M.; Teixeira, I.C.; Teixeira, J.P.; "RTL-based functional test generation for high defects coverage in digital SOCs", European Test Workshop, 2000. Proceedings, pp 99-104.

[7] Tasiran, S.; Keutzer, K., "Coverage metrics for functional validation of hardware designs", Design & Test of Computers, IEEE, Volume 18, Issue 4, July-Aug. 2001 Page(s):36 -45

[8] Fallah, F.; Devadas, S.; Keutzer, K.; "OCCOM-efficient computation of observability-based code coverage metrics for functional verification", Computer-Aided Design of Integrated Circuits and Systems, IEEE Transactions on Volume 20, Issue 8, Aug. 2001 Page(s):1003 – 1015

[9] Liang H.C., Lee C. L., Chen J.E., "Identifying untestable faults in sequential circuits". IEEE Design & Test of Computers, No. 2, 1995 pp 14-23

[10] Hsiao, M.S. "A fast, accurate, and non-statistical method for fault coverage estimation", ICCAD 1998, pp 155 – 161

[11] Akers, S.B.; Park, S.; Krishnamurthy, B.; Swaminathan, A.; "Why is less information from logic simulation more useful in fault simulation?", International Test Conference 1990.

Test Calculation for Logic and Delay Faults in Digital Circuits

József Sziray

Department of Informatics,
Széchenyi University, Győr, Hungary
E-mail: sziray@sze.hu

Abstract: The paper presents a test calculation principle which serves for producing tests for logic and delay faults in digital circuits. Switch-level logic faults in CMOS circuits are also considered. The delay faults manifest themselves in the incorrect timing behavior of some logic elements within the network. Both single and multiple faults are included. The proposed method handles multi-valued logic, where the number of logic values is unlimited. The level of circuit modeling is also allowed to vary in a wide range: switch level, gate level, functional level, register-transfer level are equally allowed. Both combinational and sequential circuits are considered. The principle is comparatively simple, and it yields an opportunity to be realized by an efficient computer program.

Keywords: *Test-pattern calculation, logic faults, CMOS transistor structures, delay faults, multi-valued logic, functional testing.*

1. Introduction

The steady development in the designing and manufacturing of digital circuits implies new fault models which require new methods in test design. The fulfillment of this requirement is especially important in the field of VLSI CMOS circuits which are rapidly spreading in the modern hardware construction. It is necessary to use efficient tests that are capable of detecting a wide range of possible faults. As known, not only permanent hardware faults are to be detected, but also intermittent faults. The detection of these faults is done during normal system operation by performing built-in self testing (BIST). At the same time, statistical analyses prove that incorrect timing behavior also plays a very significant role in hardware malfunctions.

This paper is intended to present a general test calculation principle that is suitable for producing tests for logic and timing (single, multiple) faults. The principle handles multi-valued logic, where the number of logic values is not limited. All of those representations are accepted here which operate in the logic domain by applying multi-valued logic. For instance, if the circuit behavior is described by the **VHDL language** [1], the logic values will correspond to the signal values in the actual description. Moreover, the level of circuit modeling is also allowed to vary in a wide range: switch level, gate level, functional level, register-transfer

level (RTL) are equally allowed. At the same time, the proposed principle is comparatively simple, and it yields an opportunity to be realized by an efficient computer program, both for combinational and for sequential circuits.

2. The basic principle

The test calculation principle to be presented enables the user to model the digital circuits at various levels, namely:

- **Switch level for MOS circuits:** the building elements are transistors.
- **Gate level:** logic gates are applied exclusively.
- **Functional level:** the logic values of an element are calculated with knowledge of its external functional behavior. For this modeling purpose, high level **hardware-description languages (HDL's)** can be applied, for instance, VHDL.
- **Register-transfer level:** the behavior of the digital system is described by means of bit vectors that are processed and transferred among various building blocks.

Let the vector of primary input and output variables for the complete network be $\bar{x} = (x_1, x_2,..., x_n)$ and $\bar{z} = (z_1, z_2,..., z_m)$, respectively. Let the set of possible logic values in the network be $V = \{v_1, v_2,..., v_s\}$. In addition to the elements of V, the indifferent or don't care value d will be applied.

The principle of test calculation is based on the formerly elaborated **composite justification algorithm** that has originally been intended for various permanent logic faults, such as stuck-at, bridging, and functional faults [2]-[4]. At first, the overall concept of composite justification will be summarized.

Suppose that a sequence of primary input vectors $X(t) = \bar{x}_1, \bar{x}_2,..., \bar{x}_t$ detects $q \geq 1$ simultaneous logic faults at a primary output z_j. Now the task of calculating $X(t)$ can be stated in the following way. Find a sequence of input patterns which implies $z_j = \alpha$ in the fault-free network, and $z_j = \beta$ in the faulty network, where

$$\alpha \in V, \quad \beta \in V, \text{ and } \alpha \neq \beta.$$

To reach this goal, we associate the logic values α and β with z_j, and attempt to find a sequence of input patterns which equally justifies

1) the normal value of z_j for the normal (fault-free) network, and
2) the faulty value of z_j for the faulty network.

In the first case, X(t) justifies $z_j = \alpha$ through the values of all the necessary network lines in the usual manner. In the second case, however, since the faulty values are self-dependent, they need not be justified by X(t). Thus, in the faulty network, X(t) and the faulty values justify $z_j = \beta$ jointly.

The test sequence can be derived by applying the **line-value justification concept**. As known, line-value justification is a procedure with the aim of successively assigning input values to the logic elements in such a way that they are consistent with each previously assigned value. (This concept is an auxiliary calculation process for justifying an initial set of logic values in a network, first applied in the D-algorithm for two-valued logic [5], [6].)

In our approach, the computations are carried out simultaneously in the normal and faulty network, i.e., in the normal and the faulty domain. Logic values simultaneously representing signal values in both the normal and the faulty networks are called **composite values**. Line justification performed in terms of composite values is referred to as **composite justification** [2], [3]. The two components of a composite value will be separated by a slash, with the normal component preceding the faulty one. The actual logic value of the i-th line in the network will be denoted by $v(i)$. Then, for example, a composite value of line i is

$$v(i) = v_i / v_j.$$

In the composite justification the computational costs can be greatly reduced by the following consideration [2]. The signal values in the normal and faulty networks cannot differ at the lines that do not carry any signals propagating from the sites of the faults. These are called **inactive lines**, for which $v_i \neq v_j$ would represent inconsistency if they had the composite value of v_i / v_j. It should be realized that for our purposes it is sufficient to determine which lines (called **potentially active lines**) carry signals from the faulty lines to z_j. This holds true, since all the other lines in the network are either inactive or are not involved in the justification process. The set of potentially active lines can be generated by intersecting the set of the lines that are reached from the faulty lines with the set of the lines from which z_j can be reached. The two sets are very easy to obtain by topologically tracing out the signal connections. This is done by starting from the faulty lines and proceeding forward, then starting from z_j, and proceeding

backward. If a line is encountered in both the forward and backward tracing then it is potentially active. At the end of the calculations, those lines which actually carry the fault signal to z_j will be the real **active lines**. These lines have the so-called **active composite values** which are equivalent with the **fault signals**.

The other consideration relates to the initial values associated with line z_j. It is not known in advance which normal and faulty values are to be assumed. Therefore we make an arbitrary choice. However, in the case of single faults, there is no need to repeat the justification process with interchanged values for z_j, even if the initial choice has failed. Whenever the last in a series of active values along a path between the fault site i and z_j encounters contradiction at i, we have to interchange the components of each composite value, then proceed with the calculations in the same way as before.

The implementation of the above principle for a synchronous sequential network may require the justification process to be performed through different storage states of the network, which results in a test sequence X(t) of length t. The detailed principle of doing it for stuck-at-0 / 1 faults is described in [2], [3].

In the following, we are going to present the necessary and sufficient conditions for performing the composite justification in case of various fault models.

3. Tests for stuck-at-constant logic faults

A stuck-at fault occurs when an arbitrary signal is erroneously preserving a certain logic value (for example, logic 0 or 1, high impedance value, etc.), independent of the actual control values which may influence the normal value of this signal. Any logic values are allowed within the range of possible values for that particular signal. It should be added that not only single faults, but also multiple faults are included, with no limits in their multiplicity.

Let the set of lines with stuck faults be denoted by SL, where the number of lines belonging to SL is q. If the stuck value at line i of SL is s_i, then the initial set of the logic values that are to be justified will be as follows:

$z_j = \alpha / \beta$ for a selected primary output, where

$$\alpha \in V, \quad \beta \in V, \quad \text{and} \quad \alpha \neq \beta,$$
and
$$v(i) = d / s_i \quad \text{for each} \quad i \in SL.$$

In the justification process, the value d need not be justified, whereas the stuck values s_i must not be justified.

In the following we introduce three more notations:

1550-4093/07 $25.00 © 2007 IEEE

Let the **stuck-at-α** fault of the **i-th line** be denoted by **i(α)**. The apostrophe sign will stand for logic inversion. Furthermore, the network path containing the series of lines i, j, ..., p will be denoted by

$$P(i - j - \ldots - p).$$

As an example, consider the network shown in Figure 1 where a test is to be calculated for fault **7(0)**. The potentially active lines are marked heavy in the schematic. Since an inconsistency occurred in terms of an active value at the gate with output line 10, interchange is performed. The figure indicates the logic values assigned to the lines at the current stage of the process. Figure 2 shows the values in the network upon completion of the process after the interchange.

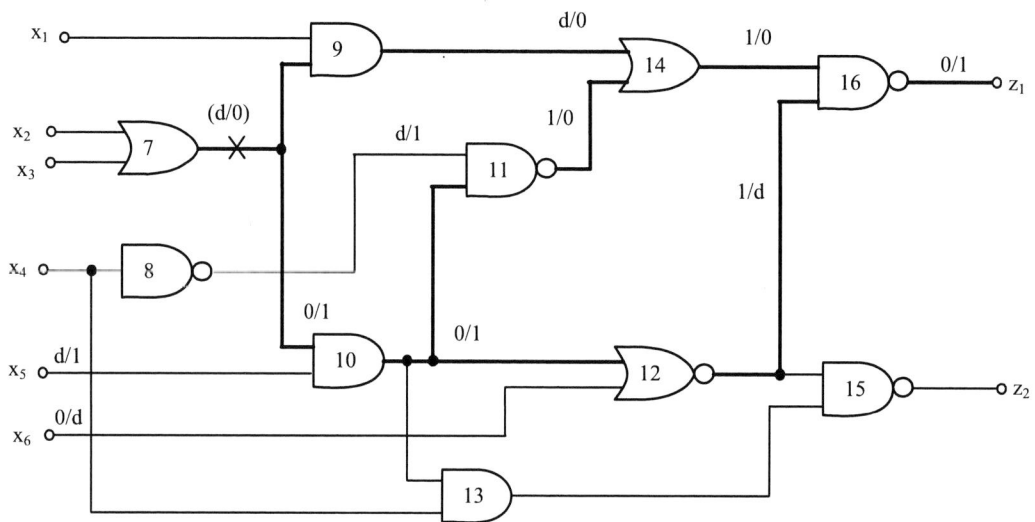

Figure 1. Composite justification before interchange.

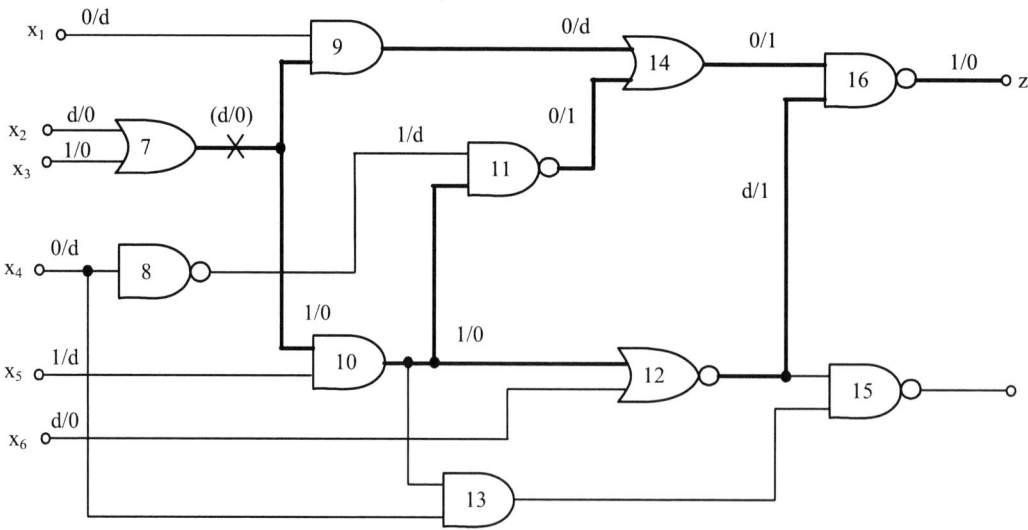

Figure 2. Composite justification after interchange.

1550-4093/07 $25.00 © 2007 IEEE

The obtained test consists of the normal components of the final composite values at the primary input lines:

$$\overline{x} = (0, d, 1, 0, 1, d).$$

As it can be seen, the test simultaneously sensitizes paths P(7-10-11-14-16) and P(7-10-12-16), while it leaves path P(7-9-14-16) necessarily unsensitized. The sensitized paths propagate the faulty signals towards the primary output z_1.

If the number of faults is greater than one, the interchange process cannot always be accomplished because of the following. When an active value reaches a faulty line, then there may be other faulty lines which have been processed and have obtained a composite value. In this case the values in the first domain cannot be generally replaced by those in the second domain. Consequently, the composite justification for multiple faults needs to be performed at first for $z_j = 0 / 1$, and if it fails then repetition for $z_j = 1 / 0$ is required.

4. Tests for CMOS transistor faults

4.1. Switch-level faults

The considered fault model of MOS circuits include stuck-at-0/1 logic faults on the connecting lines of MOS cells, and switch faults in the transistors: stuck open (the transistor is erroneously in an open circuit state), and stuck short (the transistor is erroneously in a short circuit state). The paper [7] describes a procedure to convert a transistor structure into an equivalent logic gate structure. In this way the transistor faults can be converted to stuck faults in the logic structure. All this makes it possible to use composite justification for single and multiple faults in MOS circuits, by applying some necessary modifications. The logic modeling of various MOS-circuit faults has been thoroughly discussed in [8]-[10].

Figure 3 shows an NMOS and a PMOS transistor. The transistors are controlled by their gate input. Depending upon the state (logic 0 or 1) of the gate input signal, the corresponding transistor establishes an open circuit or a short circuit between S (source) and D (drain). In the case of an NMOS transistor, the logic value 1 results in a short circuit, while the logic value 0 results in an open circuit. A PMOS transistor behaves in a complementary way.

It is easy to see that the stuck-at logic faults at the gate input of a transistor are equivalent with the short / open faults of the transistor. For example, stuck-at-1 at the gate input of an NMOS transistor is equivalent with the short-circuit fault of the transistor. This kind of equivalence enables us to use a unified fault model, namely, the tests are to be

calculated only for one type of faults, let us say, for short/open transistors.

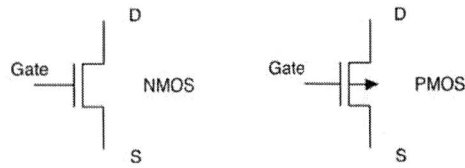

Figure 3. Transistor schemes.

In Figure 4 an NMOS type cell is presented. On the upper part of the circuit a depletion load transistor is placed. This load transistor generates a logic 1 at the output F if an open circuit exists between VSS and F. In this case F' = 0. If there is a conducting (short circuit) path between VSS and F, then F = 0 will hold.

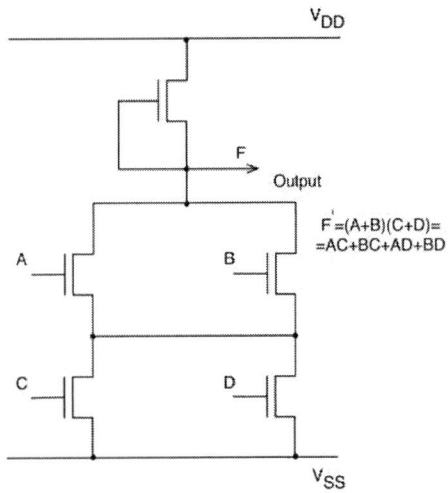

Figure 4. An NMOS cell.

4.2. Application of composite justification

The logic model for MOS transistor structures can be deduced by using conventional logic gates, AND, OR, NOT, NAND, NOR, XOR, and one additional modeling block [7], [10]. The additional block is necessary for handling the high impedance state (Z), and for eliminating VDD and VSS from the logic model [10]. This modeling block will be referred to as block "B". It has two inputs, S0 and S1, and one output F.

S0 represents the switch function for the connection towards VSS. S1 represents the switch function for the connection towards VDD. Each of these functions is logic 1 if the switch is shorted, and it is logic 0 if the switch is open. The switch is shorted if there is at least one short-circuit path towards VSS or VDD. The switch is open if there exists no short-circuit path towards VSS or VDD. The truth table of block B is defined below:

S0	S1	F
0	0	M
0	1	1
1	0	0
1	1	0

Table 1. The logic behavior of block B.

When both inputs are 0, the output is M. Here M represents the memory or the previous state (before S0 and S1 changed to 0) of the output line. M can be any value among 0, 1, or "unknown". When both inputs are 1, the output is shown as 0. This value is taken under the assumption that the path from VSS dominates over the path from VDD, and hence the output is pulled to the logic value of 0. If the assumption is not true for any particular

technology, this value can be set to 1 or "unknown", as necessary.

The extension of composite justification to CMOS circuits requires only the capability of handling the block B. It can be done in a straightforward way, on the basis of the truth table.

When considering the complete CMOS circuit, the initial setting is

$$z_j = \alpha / \beta \text{ for a selected primary output, where } \alpha \neq \beta.$$

In our case, the possible values for α and β are logic 0 and 1, as well as high impedance state (Z).

The calculation procedure will be illustrated by an example selected from [7]. The CMOS circuit in Figure 5 serves for this purpose. The equivalent logic model of the circuit is shown in Figure 6. Here within a block B terminal S0 is represented by a logic 0, while terminal S1 is represented by a logic 1.

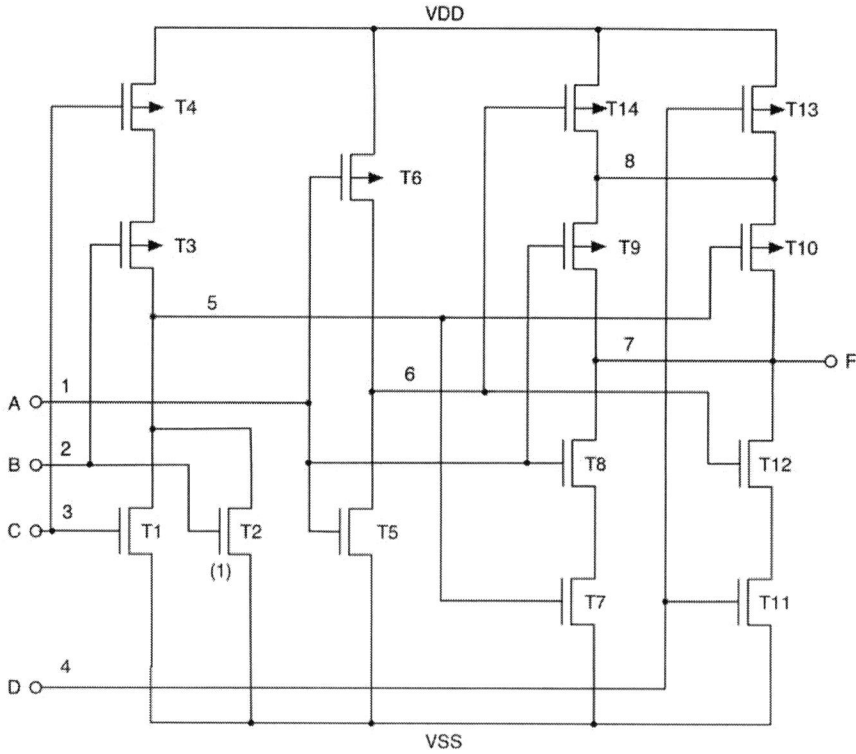

Figure 5. CMOS circuit for test calculation.

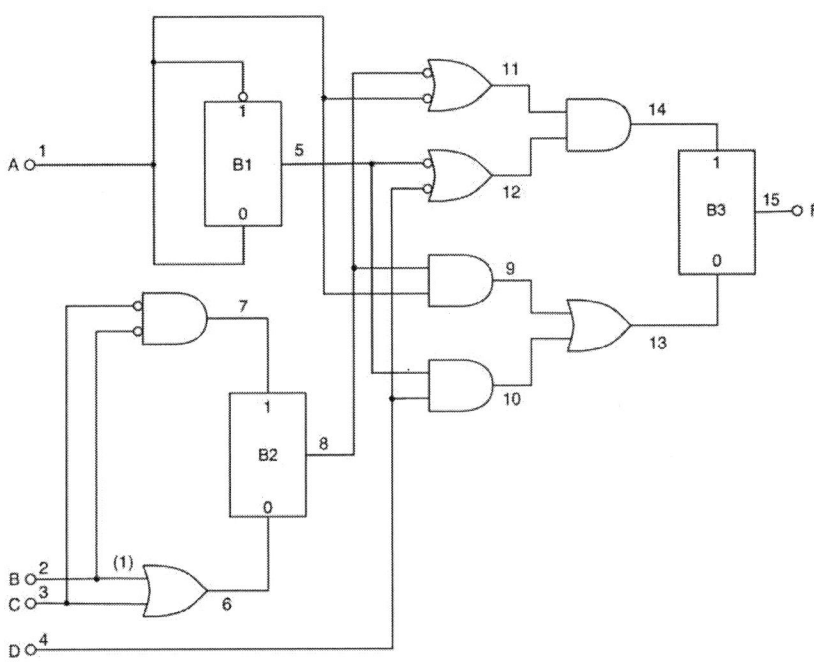

Figure 6. Logic model for the circuit of Fig. 5.

Suppose that the fault in the circuit is stuck short of the transistor T2. As known, this fault is equivalent with the stuck-at-1 fault at the gate input of T2, i. e., with fault 2(1). The calculation process is carried out in the following way:

First the set of potentially active lines is determined. It consists of the elements listed below:

$$\{2, 6, 7, 8, 9, 11, 13, 14, 15\}.$$

Next the consecutive steps of line-value justification are presented:

Step 1: $F = 0 / 1$, $z1 = v(15) = 0 / 1$.

Step 2: $v(14) = 0 / 1$, $v(13) = 1 / 0$.

Step 3: $v(11) = 0 / 1$, $v(12) = d / 1$.

Step 4: $v(9) = 1 / 0$, $v(10) = d / 0$.

Step 5: $v(8) = 1 / 0$, $v(1) = 1 / d$.

Step 6: $v(7) = 1 / 0$, $v(6) = 0 / 1$.

Step 7: $v(2) = 0 / 1$, $v(3) = 0 / d$.

The obtained test consists of the normal components of the final composite values at the primary input lines:

$$A = 1, \quad B = 0, \quad C = 0, \quad D = d.$$

The same test in vector form is

$$\overline{X}_t \ \ = (1, 0, 0, d).$$

As it can be seen, the test simultaneously sensitizes paths P(2-7-8-11-14-15) and P(2-6-8-9-13-15). The sensitized paths propagate the faulty signals towards the primary output F.

5. The calculation process for delay faults

5.1. The fault model

Incorrect timing behavior of a logic network is due to the improper delays of certain logic elements [11]-[14]. In order for the circuit to operate in the right way it is necessary that each element have its propagation-time delay within a specified minimum and a maximum value. A **delay fault** of a logic element is defined as follows: A fault is said to be **negative** if the delay is less than the specified minimum time. Similarly, a fault is **positive** if the delay is more than the specified maximum time. In the following, we are going to deal with positive delay faults, where combinational networks are considered. Since there is no principal difference between the two types of faults, methods for detecting a positive delay are readily extendable to negative delays. At the same time, the results for a combinational network can also be extended to synchronous sequential networks.

1550-4093/07 $25.00 © 2007 IEEE   25

In the case of delay faults, we will assume that the overall static behavior of the complete network is correct, i.e., no logic faults are encountered.

The faults are assumed at the output nodes of the logic elements. Two fault types will be taken into consideration:

- Delay in **rising transition**, where the rising edge of the output signal is out of time.
- Delay in **falling transition**, where the falling edge of the output signal is out of time.

For the sake of brevity, these two types of faults will be referred to as **rising delay**, and **falling delay**, respectively.

A test of a delay fault consists of two consecutive input vectors, $\bar{x}_1$ and $\bar{x}_2$ which differ in at least one bit. The first vector $\bar{x}_1$ is loaded to the circuit under test at time t_1. After the signals in the circuit have stabilized, the second input vector $\bar{x}_2$ is loaded at time t_2. By sampling the output signal of z_j at time t_3, where $(t_3 - t_2)$ corresponds to the desired operational time interval, one can determine the existence of the delay fault. A delay fault of an element is detected at a primary output z_j, if and only if the implied signal transition at z_j does not takes place within the time interval $(t_3 - t_2)$. The required time interval is calculated as the sum of the delays at the elements along a signal transition path from a primary input to the primary output z_j. The correct and the erroneous behavior are illustrated in Figure 7 and Figure 8, respectively.

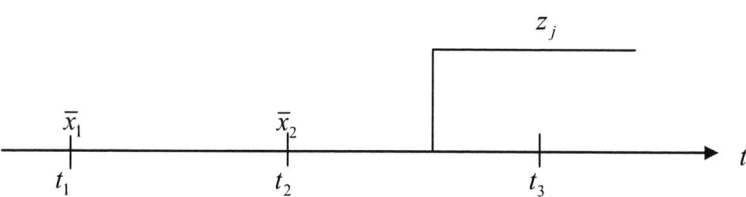

Figure 7. Correct timing behavior.

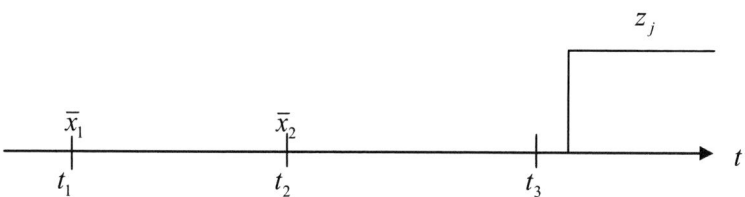

Figure 8. Erroneous timing behavior.

Although the composite justification has been elaborated for logic faults, its concept can be extended to timing faults in a straightforward manner. The appropriate solution is as follows:

In the test calculation process, the two components of a composite value correspond to the values in the network taken in the presence of the two components of the test itself. It means that the first component belongs to the first input vector of the test, while the second component belongs to the second input vector. Here the calculations are to be performed in two domains again, where the first domain represents the network when $\bar{x}_1$ is applied, while the second domain is for $\bar{x}_2$. Of course, no logic faults are involved, i.e., the static network behavior is supposed to be correct in both domains.

Let the set of lines with delay faults be denoted by DL, where the number of lines belonging to DL is $k \geq 1$. If we first assume that two-valued logic is used in the network, i.e., $V = \{0, 1\}$, the initial set of the logic values that are to be justified will be

$$z_j = \alpha \, / \, \alpha'$$

for a selected primary output, where $\alpha \in \{0, 1\}$. Furthermore,

   $v(i) = d \, / \, 1$ in case of a rising fault, and
   $v(i) = d \, / \, 0$ in case of a falling fault,
for each $i \in DL$.

In the justification process the inactive and potentially active lines are also to be taken into account. An inactive line is allowed to have the same logic values for the two test vectors. A potentially active line has the possibility to propagate a falling (1 / 0) or a rising (0 / 1) signal transition towards z_j. It should be noted that these transition signals directly correspond to the fault signals that are treated for logic faults. Consequently, the propagation of transition signals is the same as the propagation of fault signals, towards a primary output selected for fault detection.

5.2. A calculation example

The above test calculation concept is illustrated in this section of the paper. In order to do that, first we introduce the following notation.

The delay fault of a logic element is assumed to appear at an output line of that element. Let the rising fault of the i-th line be denoted by **i(0/1)**, and the falling fault of it by **i(1/0)**.

As an example, consider again the network shown in Figure 1 where a test is to be calculated for fault **7(1/0)**. In the case of this particular fault, exactly the same computations will be performed as for the single fault **7(0)**. It means that the interchange step is also to be carried out. Figure 2 shows the values in the network upon completion of the process after the interchange.

The obtained test consists of the following two consecutive input vectors:

$$\overline{x}_1 = (0, d, 1, 0, 1, d), \qquad \overline{x}_2 = (d, 0, 0, d, d, 0).$$

Now the test simultaneously sensitizes paths P(7-10-11-14-16) and P(7-10-12-16), while it leaves path P(7-9-14-16) necessarily unsensitized. The sensitized paths propagate the transition signals towards the primary output z_1.

5.3. Further extensions

As far as the number of logic values is concerned, it can easily be seen that the above principle is not limited to only two values. In the case of **multi-valued logic**, we only have to handle signal transitions between any two different values.

In this case, if the final transition value of the delay fault at line i of DL is s_i, then the initial set of the logic values that are to be justified will be as follows:

$$z_j = \alpha / \beta \text{ for a selected primary output, where}$$
$$\alpha \in V, \quad \beta \in V, \text{ and } \alpha \neq \beta, \text{ and}$$
$$v(i) = d / s_i \text{ for each } i \in DL.$$

Finally, it should be added that there exists another widely spread fault model, the so-called **path-delay fault** [12]-[14]. These faults manifest themselves in cumulative propagation delays along network paths, where the path delay is under or above the specified range. Rising and falling delays may also be distinguished for a complete path. The papers [13] and [14] present test calculation procedures for such faults by using a multi-valued

logic system. The composite justification method can also be extended to handling paths instead of elements. In this approach, assuming two-valued logic, the initial values of the start line x_i and the end line z_j of the path will be

$$x_i = \alpha / \alpha', \qquad \text{and} \qquad z_j = \beta / \beta',$$

where $\alpha \in \{0, 1\}$, $\beta \in \{0, 1\}$.

In case that the actual path is located in a gate level network the values of α and β can be determined in advance. If the number of logic inversions along the whole path is even, the values of α and β will be equal. If the number of inversions is odd, the two logic values must be different. For the rising fault of the path, $z_j = 0 / 1$, and for the falling fault $z_j = 1 / 0$.

If the building elements of a logic network are **multiple-output modules**, the transition delay for a module can be treated in the following way: Exact modeling requires that each module output have its own falling/rising delays which may be different from those of the other outputs. However, this situation does not necessitate any changes in the calculation process, since the module outputs are to be taken into account as if they were different gate outputs.

6. Conclusions

This paper has been meant for showing how the test calculation algorithm first published in [2] can be generalized for treating logic and delay faults in a wide range of circuit models. The flexibility of the composite justification is based on the fact that it establishes **the minimal necessary and sufficient set of logic values** which yield the test conditions for the faults. As seen, the tests are obtained by justifying the initial logic conditions. The salient advantage of composite justification is **the total absence of the fault propagation** phase. This feature makes the approach extremely flexible in terms of circuit modeling and fault classes. The same applies to the use of an HDL. As known, the fault propagation phase, i.e., D-propagation is an inherent part of the wide-spread D-algorithm, where this phase implies serious difficulties for functional level models and multiple faults. Functional algorithms for constructing computational tools of complex logic modules have been presented in [15]. This paper clearly illustrates the problems encountered in this topic.

When multiple faults are considered, the complete procedure of the D-algorithm has to be repeated $2^q - 1$ times in worst case, for one primary output, where q is the multiplicity of faults. In this case, attempts are made to propagate different combinations of the individual faults. On the other hand, composite justification requires only 2 iterations in worst case, also for one primary output.

1550-4093/07 $25.00 © 2007 IEEE

There exist other test generation algorithms like PODEM [16] and FAN [17] that have been proved to be more efficient than the D-algorithm. These solutions can also be extended to treating delay faults, e.g., PODEM has been utilized in [13]. As a matter of fact, the acceleration results in the PODEM and FAN are relied on the use of gate-level network structure. However, if functional level modeling is considered then the structural and logic analyses performed in both algorithms become extremely cumbersome and hardly feasible. The basic results achieved in the area of treating delay faults are also related to gate-level descriptions. In contrast, the approaches based on composite justification are not really limited by the way of network modeling. The same applies to the network types, i.e., combinational or sequential, and also, the types and multiplicity of the faults.

As far as the computer implementation is concerned, only line justification is to be accomplished in the presented principle, which is also an advantage. In order to perform the line-value justification, the **inverse models** of the building elements in the network are required. An inverse model defines the set of possible input patterns which result in a specific state or an output pattern [15], [18]. For this purpose, high level hardware-description languages, such as VHDL can also be applied [19], [20].

The computerized implementation of the proposed principle will handle multi-valued logic, where the number of logic values correspond to the number of the possible signal values in the HDL applied for modeling.

The complexity of a logic module is not limited in principle. Only practical trade-offs, such as processing time, and user efforts in terms of modeling accuracy are to be considered.

Finally, it should be added that in terms of delay faults, the proposed calculation process is intended above all for element faults, instead of path faults. It is able to produce a test for each detectable fault of any element in the network. On the other hand, those approaches which deal with a selected path [12]-[14] do not yield the same performance. The reason is as follows: They are aimed at sensitizing only one path at a time. If it fails then a new path is selected, etc. However, due to **path reconvergence**, it may well occur that more than one paths have to be sensitized simultaneously, which is not involved in these approaches. An example for this situation was shown in Figure 2.

References

[1] Z. Navabi, "VHDL: Analysis and Modeling of Digital Systems", McGraw-Hill, Inc., USA, 1993.

[2] J. Sziray, "Test calculation for logic networks by composite justification", Digital Processes, Vol. 5, No. 1-2, pp. 3-15, 1979.

[3] J. Sziray, "Functional level test calculation and fault simulation for logic networks", Discrete Simulation and Related Fields (Ed. by A. Jávor), pp. 223-234, North-Holland Publishing Company, Amsterdam, 1982.

[4] J. Sziray, "A comprehensive method for the test calculation of complex digital circuits", Periodica Polytechnica, Budapest Technical University, Series of Electronic Engineering, Vol. 41, No. 4, pp. 251-257, 1998.

[5] J. P. Roth, "Diagnosis of automata failures: a calculation and a method", IBM Journal of Research and Development, Vol. 10, pp. 278-291, July 1966.

[6] M. Abramovici, M. A. Breuer, A. D. Friedman, "Digital Systems Testing and Testable Design", Computer Science Press, USA, 1990.

[7] S. K. Jain, V. D. Agrawal, "Modeling and test generation algorithms for MOS circuits", IEEE Trans. on Computers, Vol. C-34, pp. 426-433, May 1985.

[8] R. L. Wadsack, "Fault modeling and logic simulation of CMOS and MOS integrated circuits", Bell System Technical Journal, Vol. 57, No. 5, pp. 1449-1473, May-June 1978.

[9] J. Galiay, Y. Crouzet, M. Vergniault, "Physical versus logical fault models in MOS LSI circuits, inpact on their testability", The Ninth Annual International Symposium on Fault-Tolerant Computing (FTCS-9), pp.195-202, Madison, Wisconsin, June 20-22, 1979.

[10] Y. H. Levendel, P. R. Menon, and C. E. Miller, "Accurate logic simulation models for TTL totem-pole and MOS gates and tristate devices", Bell System Technical Journal, Vol. 60, pp. 1271-1287, September 1981.

[11] R. B. Hitchcock, G. L. Smith, D. D. Cheng "Timing analysis of computer hardware", IBM Journal of Research and Development, Vol. 26, pp. 100-105, January 1982.

[12] J. P. Lesser, J. J. Shedletsky, "An experimental delay test generator for LSI logic", IEEE Trans. on Computers, Vol. C-29, pp. 235-248, March 1980.

[13] C. J. Lin, S. M. Reddy, "On delay fault testing in logic circuits", IEEE Trans. on Computer-Aided Design, Vol. CAD-6, pp. 694-703, September 1987.

[14] I. Pomeranz, S. M. Reddy, "A generalized test generation procedure for path delay faults", FTCS-28, International Symposium on Fault-Tolerant Computing, Proceedings, pp. 274-283, Munich, June 1998.

[15] M. A. Breuer, A. D. Friedman, "Functional level primitives in test generation", IEEE Trans. on Computers, Vol. C-29, pp. 223-235, March 1980.

[16] P. Goel, "An implicit enumeration algorithm to generate tests for combinational logic circuits", IEEE Trans. on Computers, Vol. C-30, pp. 215-222, March 1981.

[17] H. Fujiwara, T. Shimono, "On the acceleration of test generation algorithms", IEEE Trans. on Computers, Vol. C-32, pp. 1137-1144, December 1983.

[18] J. Sziray, Zs. Nagy, "OPART: A hardware-description language for test generation", Microprocessing and Microprogramming, (Ed. by P. Nunez), pp. 525-530, North-Holland Publishing Company, Amsterdam, 1991.

[19] B. Sallay, A. Petri, K. Tilly, A. Pataricza, B. Benyó, J. Sziray, "High level test pattern generation for VHDL circuits", IEEE European Test Workshop '96, Proceedings, pp. 201-205, Montpellier, June 1996.

[20] B. Benyó, J. Sziray, "The use of VHDL models for design verification", IEEE European Test Workshop, Proceedings, pp. 289-290, Cascais, Portugal, May 23-26, 2000.

SECTION 2:
VERIFICATION AND DESIGN ISSUES

Directed Micro-architectural Test Generation for an Industrial Processor: A Case Study

Heon-Mo Koo   Prabhat Mishra
Computer and Information Science and Engineering
University of Florida, Gainesville, FL 32611, USA.
{hkoo, prabhat}@cise.ufl.edu

Jayanta Bhadra   Magdy Abadir
Freescale Semiconductor Inc.
7700 West Parmer Lane Austin, TX 78727, USA
{Jayanta.Bhadra, M.Abadir}@freescale.com

Abstract

Simulation-based validation of the current industrial processors typically use huge number of test programs generated at instruction set architecture (ISA) level. However, architectural test generation techniques have limitations in terms of exercising intricate micro-architectural artifacts. Therefore, it is necessary to use micro-architectural details during test generation. Furthermore, there is a lack of automated techniques for directed test generation targeting micro-architectural faults. To address these challenges, we present a directed test generation technique at micro-architectural level for functional validation of microprocessors. A processor model is described in a temporal specification language at micro-architecture level. The desired behaviors of micro-architecture mechanisms are expressed as temporal logic properties. We use decompositional model checking for systematic test generation. Our experiments using a processor based on the Power Architecture^{TM} Technology[1] shows very promising results in terms of test generation time as well as test program length.

1 Introduction

Performance improvement of modern processors is accompanied with high design complexity by adopting complicated micro-architectural mechanisms such as deeply pipelined superscalar, dynamic scheduling, and dynamic speculation. Since verification complexity is directly proportional to the design complexity, considerable amount of time and resources are spent on design verification.

In the current industrial practice [11], random and biased-random test generation techniques at architecture (ISA) level are most widely used for simulation-based validation to uncover errors early in the design cycle as well as to perform simulation for the entire processor design. However, as demonstrated in Section 4, architectural test generation techniques have difficulty in activating micro-architectural target artifacts and pipeline functionalities since it is not possible to generate information regarding pipeline interactions or timing details using input ISA specification. For example, it is very hard to

generate an architectural test program for micro-architectural design bugs such as a pipeline interaction error (e.g., "decode stage is not stalled even if Completion Queue is full"), or a performance error (e.g., "data dependency is not resolved by forwarding path even if operand is available"). Therefore, it is necessary to use micro-architectural details during test generation.

Compared to the random or biased-random tests, the directed tests can reduce overall validation effort since shorter tests can obtain the same coverage goal. However, there is a lack of automated techniques for directed test generation targeting micro-architectural faults. As a result, directed tests are hand-written by experts. Due to manual development this process can be error prone.

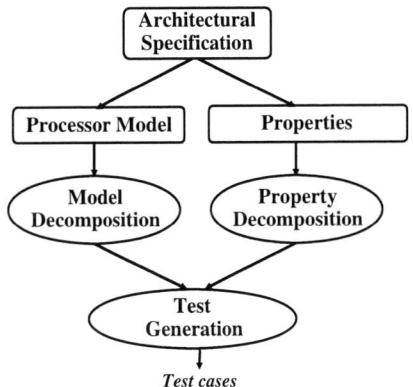

Figure 1. Test Program Generation Methodology

To address these challenges, we present a directed test generation technique at micro-architectural level for functional validation of microprocessors. Figure 1 shows our directed test generation methodology. The input specification contains both the structure (micro-architectural details) and the behavior (instruction-set) of the processor. The micro-architectural features in the processor model include pipelined and clock-accurate behaviors that enable micro-architectural test generation. Properties can be automatically generated from the input specification based on a functional fault model such as pipeline (graph) coverage [9]. Additional properties can be added based on interesting scenarios such as combined pipeline stage rules and corner cases. These properties are described in temporal

---

[1]The Power Architecture and Power.org wordmarks and the Power and Power.org logos and related marks are trademarks and service marks licensed by Power.org

1550-4093/07 $25.00 © 2007 IEEE

logic. For automatic test generation, we use decompositional model checking where the processor model as well as the properties are decomposed and the model checking is applied on smaller partitions of the design using decomposed properties. We introduce the notion of time steps to enable decomposition of the properties into smaller ones based on their clock cycles. We have developed an efficient algorithm to merge the partial counterexamples generated by the decomposed properties to produce the global counterexample corresponding to the original property. We applied this methodology on a processor based on the Power Architecture Technology to demonstrate the usefulness of our approach.

The main contribution of this work is to establish a framework for a directed and automated micro-architectural test generation technique for validation of modern industrial processors. Since the proposed method is generic, its framework can be used for validation of any other real processors. The rest of the paper is organized as follows. Section 2 presents related work addressing test generation in the context of micro-architectural validation of pipelined processors. Section 3 describes our test generation methodology. Section 4 presents a case study and Section 5 concludes the paper.

2 Related Work

Several methodologies have been developed for validation of pipelined processors using finite state machine (FSM) models [1, 5, 6, 10] where FSM coverage is used to generate test programs. In modern processor designs, complicated micro-architectural mechanisms include interactions among many pipeline stages and buffers that lead the FSM-based approaches to the state-space explosion problem. To alleviate the state explosion, Utamaphethai et al. [12] have presented a FSM model partitioning technique based on micro-architectural pipeline storage buffers whose entries store data and status. However, it suffers from targeting complete micro-architectural features because test programs are generated by design errors from each buffer, not for combined buffers.

An alternative formal method, model checking [2], has been successfully used in software and hardware validation as a test generation engine [3, 9]. The negated version of a desired property along with the processor model is applied to the model checker. The model checker automatically produce a counterexample that contains a sequence of instructions (a test program) from an initial state to a failure state. However, this naive approach is unsuitable for a real processor model due to the state explosion problem during model checking.

Koo and Mishra [7, 8] have proposed a processor/property decomposition technique to reduce the search space during counterexample generation as well as an algorithm for merging the partial counterexamples to generate architectural test programs. Their test generation technique is built on a relatively simple MIPS processor [4] with no renaming buffer, reordering buffer, or reservation station. They use pipeline path-level model partitioning to generate a test program for data forwarding, but it causes deprivation of memory during model checking

when applying to the industrial processors due to high complexity of even single pipeline path. In addition, they mainly focus on the data path rather than the control path. While a data (opcode and operands) is located in a single pipeline stage, control signals (functional unit status and buffer status) may spread across multiple pipeline stages and buffers which make model partitioning and counterexample merging more difficult. Therefore, it is necessary to improve the decomposition and merging algorithms for application to the complex industrial processors.

3 Directed Micro-architectural Test Generation

Today's test generation techniques and verification methods are very efficient to find bugs at the unit level. Hard-to-find bugs arise often from the interactions among many pipeline stages and buffers of a modern processor design. We primarily focus on such micro-architectural interface among functional units in a pipelined processor.

Algorithm 1: *Test Generation*
Inputs: i) Processor model M
         ii) Set of interactions S from fault model and corner cases
Outputs: Test programs
Begin
   TestPrograms = ϕ
   for each interaction S_i in the set S
      P_i = CreateProperty(S_i)
      $\overline{P_i}$ = Negate(P_i)
      $test_i$ = DecompositionalMC($M, \overline{P_i}$)
      TestPrograms = TestPrograms $\cup\ test_i$
   endfor
   return TestPrograms
End

Algorithm 1 describes our test generation procedure. This algorithm takes the processor model M and desired pipeline interactions S as inputs and generates test programs. The processor model is described in a temporal specification language such as SMV [13]. For each interaction S_i, the algorithm produces one test program $test_i$. S_i is composed of a set of instruction and control functionalities at pipeline units and their relations and it is converted to a temporal logic property P_i. The negation of P_i is an interaction fault. The processor model M and the fault $\overline{P_i}$ are applied to decompositional model checking framework to generate a test program. The algorithm iterates over all the interaction faults in the fault model and corner cases.

3.1 Micro-architectural Modeling

Figure 2 shows a functional block diagram of the four-wide superscalar e500 processor that is based on the Power Architecture Technology [14] with the seven pipeline stages. Pipeline buffers are highlighted in grey. We have developed a processor model based on the micro-architectural structure, the instruction behavior, and the rules in each pipeline stage that determine when instructions can move to the next stage and when

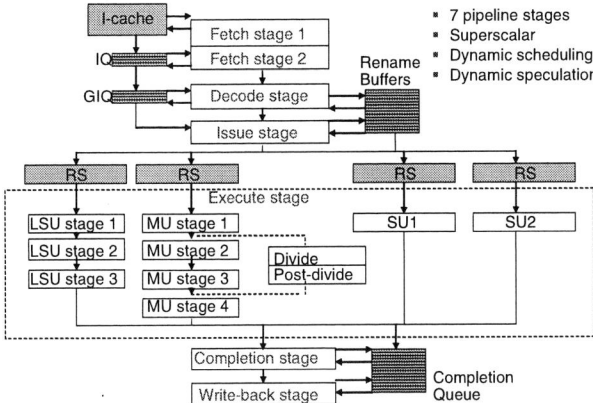

- 7 pipeline stages
- Superscalar
- Dynamic scheduling
- Dynamic speculation

Figure 2. Instruction Pipeline Flow of e500 processor that is based on the Power Architecture Technology

they cannot. The micro-architectural features in the processor model include pipelined and clock-accurate behaviors such as multiple issue for instruction parallelism, out-of-order execution and in-order-completion for dynamic scheduling, register renaming for removing false data dependency, reservation stations for avoiding stalls at Fetch and Decode pipeline stages, and data forwarding for early resolution of RAW data dependency. By representing them in a model checking language, we can achieve the automatic test generation goal.

In order to use model checking as a test generator, the processor model needs to be verified beforehand. Since it is infeasible to verify the entire model as a single unit due to the state explosion during model checking, we have partitioned the entire processor model into multiple modules based on the functional units shown as rectangles in Figure 2. Each partitioned module has been verified using the requirements and rules described in the specification. For verification of module interface, we integrated neighboring modules and verified their interface. These modules are basic units in processor model decomposition for test generation.

3.2 Property Generation and Decomposition

We generate a property for each pipeline interaction from the specification. Since interactions at a given cycle are semantically explicit and our processor model is organized in structure-oriented modules, the interactions can be converted into properties. The generated properties are expressed in LTL (Linear Temporal Logic) [2]. Each property consists of temporal operators (G, F, X, U) and Boolean connectives ($\land$, $\lor$, $\neg$, and $\rightarrow$). Most pipeline interactions can be converted in the form of a property $F(p_1 \land p_2 \land \ldots \land p_n)$ that combines activities p_i over n modules using logical *AND* operator. The atomic proposition p_i is a functional activity at a module i such as operation execution, stall, exception or NOP. The property is true if $(p_1 \land p_2 \land \ldots \land p_n)$ becomes true at any time step.

Since we are interested in counterexample generation, we need to generate the negation of the property first. The negation

of the properties can be expressed as:

$$\neg X(p) = X(\neg p) \qquad \neg G(p) = F(\neg p)$$
$$\neg F(p) = G(\neg p) \qquad \neg pRq = \neg pU\neg q$$

For example, the negation of the interaction property is $G(\neg p_1 \lor \neg p_2 \lor \ldots \lor \neg p_n)$ that becomes true if any of p_1, p_2, $\ldots$, *or* p_n is not true over all time steps. In the remainder of this section, we describe how to decompose these properties (already negated) for efficient model checking. There are various combinations of temporal operators and Boolean connectives where decompositions are not possible e.g., $F(p \land q) \neq F(p) \land F(q)$ and $G(p \lor q) \neq G(p) \lor G(q)$. In certain situations, such as pUq, $F(p \rightarrow F(q))$, or $F(p \rightarrow G(q))$, decompositions are not beneficial compared to traditional model checking. The following combinations allow simple property decompositions.

$$G(p \land q) = G(p) \land G(q) \qquad F(p \lor q) = F(p) \lor F(q)$$
$$X(p \lor q) = X(p) \lor X(q) \qquad X(p \land q) = X(p) \land X(q)$$

Introducing the notion of clock (time step) in the property allows more decompositions for counterexample generation as shown below[2]. Note that the left and right hand side of the decomposition are not logically equivalent but they produce functionally equivalent counterexamples.

$$G((clk \neq t_s) \lor (p \lor q)) \approx G((clk \neq t_s) \lor p) \lor G((clk \neq t_s) \lor q)$$

Although we only use a few decomposition scenarios, it is important to note that these scenarios are sufficient for generating the properties where interactions are considered. In addition to these interaction properties, we created many micro-architectural properties based on real experiences of industrial designers.

3.3 Test Generation using Model Checking

The basic idea of DecompositionalMC( ) in Algorithm 1 is to apply the decomposed properties (sub-properties) to appropriate modules and compose their responses to construct the final test program. Model checker is used to generate partial counterexamples for the partitioned modules. Integration of these partial counterexamples is a challenge due to the fact that the relationships among decomposed modules and sub-properties are not preserved at whole design level in general. We propose clock-based integration of partial counterexamples.

For example, if two sub-properties are applied at the same clock cycle ($clk = t_s$) to two modules sharing a parent module, then two counterexamples are generated and merged into the output property of the parent module for generating the counterexamples at the previous clock cycle ($clk = t_s - 1$). In Figure 2, four reservation station (RS) modules share the parent module Issue. Counterexamples generated from multiple RS at the cycle k are merged for creating the output property of Issue stage. The negated version of this property is applied to

---

[2]The *clk* variable is used to count time steps, and t_s is a specific time step.

1550-4093/07 $25.00 © 2007 IEEE

Table 1. Test Cases and Code Length

	Test Cases	Test Code Length
1	Instruction dual issue	15
2	Renaming *src1* operand	12
3	Read operand from forwarding path (RAW)	9
4	Reservation station reads operand from forwarding path (RAW)	7
5	Read operand from renaming reg. (RAW)	10
6	Read operand from GPR (RAW)	11

the model checker along with Issue module to generate a counterexample at the cycle $k - 1$ that is used to produce the output properties of Decode, GIQ, and Rename buffer. Merging partial counterexamples continues until we obtain the primary input assignments for all the sub-properties. These assignments contain fetched instruction data from I-cache and they are converted into assembly instruction sequences.

4 Experiments

We applied our methodology on a superscalar processor based on the Power Architecture Technology. We performed various test generation experiments for validating the pipeline interactions and corner cases. Table 1 shows a subset of the directed test cases that we generated and their corresponding length in terms of number of instruction sequences. For example, test programs for case 3 through 6 exercise operand read from four different resources as shown in Figure 3, which can be generated at micro-architecture level but very difficult at ISA level. In terms of efficiency, only several seconds were spent on test generation.

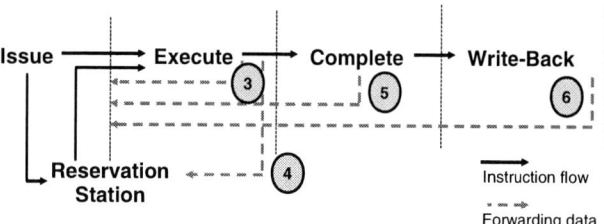

Figure 3. Four Different Data Forwarding Mechanisms

To validate these test cases, we converted the test programs into the input format of RTL simulation and monitored instructions in pipeline stages at every clock cycle during simulation to ensure that the generated test program activates the actual micro-architectural fault.

5 Conclusions

Architectural test generation techniques have limitations to achieve micro-architectural coverage goal. This paper presented a directed test generation technique based on decomposition of both processor model and properties for validation of performance as well as functionality of the modern microprocessors. Our experimental results using e500 processor that is based on the Power Architecture Technology demonstrate

the efficiency of our method by generating complicated micro-architectural tests. Since the proposed technique is generic, its framework can be used for validation of industrial-strength processors. Our future work includes extension of the processor model for dynamic speculation and other features. Since the number of interactions (directed tests) can be still extremely large, we plan to develop a test compaction technique to reduce the number of test programs.

References

[1] D. Campenhout, T. Mudge, and J. Hayes. High-level test generation for design verification of pipelined microprocessors. *DAC*, pages 185–188, 1999.

[2] E. M. Clarke, O. Grumberg, and D. A. Peled. *Model Checking.* MIT Press, Cambridge, MA, 1999.

[3] A. Gargantini and C. Heitmeyer. Using model checking to generate tests from requirements specifications. In *ACM SIGSOFT Software Engineering Notes*, volume 24, pages 146–162, 1999.

[4] J. Hennessy and D. Patterson. *Computer Architecture: A Quantitative Approach.* Morgan Kaufmann, 2002.

[5] H. Iwashita, S. Kowatari, T. Nakata, and F. Hirose. Automatic test pattern generation for pipelined processors. *ICCAD*, pages 580–583, 1994.

[6] K. Kohno and N. Matsumoto. A new verification methodology for complex pipeline behavior. *DAC*, pages 816–821, 2001.

[7] H.-M. Koo and P. Mishra. Functional test generation using property decompositions for validation of pipelined processors. *DATE*, pages 1240–1245, 2006.

[8] H.-M. Koo and P. Mishra. Test generation using SAT-based bounded model checking for validation of pipelined processors. *GLSVLSI*, 2006.

[9] P. Mishra and N. Dutt. Graph-based functional test program generation for pipelined processors. *DATE*, pages 182–187, 2004.

[10] J. Shen and J. A. Abraham. An RTL abstraction technique for processor microarchitecture validation and test generation. *Journal of Electronic Testing: Theory and Applications*, 16(1-2):67–81, 2000.

[11] K. Shimizu, S. Gupta, T. Koyama, T. Omizo, J. Abdulhafiz, L. McConville, and T. Swanson. Verification of the cell broadband engine processor. *DAC*, pages 338–343, 2006.

[12] N. Utamaphethai, R. D. S. Blanton, and J. P. Shen. Effectiveness of microarchitecture test program generation. *IEEE Design & Test*, 17(4):38–49, 2000.

[13] www-cad.eecs.berkeley.edu/ kenmcmil/smv. *Cadence SMV.*

[14] www.freescale.com/files/32bit/doc/refmanual/e500CORERMAD.pdf. *Freescale PowerPc e500 core family reference manual.*

1550-4093/07 $25.00 © 2007 IEEE

Advanced SAT-Techniques for Bounded Model Checking of Blackbox Designs *

Marc Herbstritt   Bernd Becker   Christoph Scholl

Institute of Computer Science, Albert-Ludwigs-University
79110 Freiburg im Breisgau, Germany
{herbstri,becker,scholl}@informatik.uni-freiburg.de

Abstract

In this paper we will present an optimized structural 01X-SAT-solver for bounded model checking of blackbox designs that exploits semantical knowledge regarding the node selection during SAT search. Experimental results show that exploiting the problem structure in this way speeds up the 01X-SAT-solver considerably. Additionally, we give a concise first-order formulation that is more expressive than using 01X-logic. This formulation leads to hard-to-solve QBF formulas for which experimental results from the QBF Evaluation 2006 are presented.

1 Introduction

Today's verification of circuit designs heavily makes use of property checking. As algorithmic workhorse, model checking techniques are used. Within the large pool of different approaches, SAT-based bounded model checking [1] has become an established technique in academic research as well as in industrial tools. While typically the design under analysis is fully specified, the analysis of blackbox designs emerged as an interesting problem that has a large variety of applications [2, 3, 4], e.g., early design verification, error diagnosis, etc. A blackbox design corresponds to an incomplete circuit description, i.e., some parts of the circuits are not known.

In [5] we have proposed a SAT-based approach for bounded invariant checking of such blackbox designs. Since pure propositional logic is not expressive enough for the analysis of blackbox designs, one has to apply either logical extensions like 01X-logic or first-order logic where universal and existential quantifiers allow for a concise problem formulation. The usage of 01X-logic was analyzed

already in [5]. There, 01X-logic was algorithmically integrated into a structural SAT-solver. Then it was compared to an encoding approach in the style of [2]. As a result we have found that generally the encoding approach is much faster, although in some cases the structural 01X-SAT-solver performed better. The reason for that is the "blindness" of the encoding approach regarding the correspondence of And-Inverter-Graph (AIG) vertices that together build a semantical unit within 01X-logic.

In this work, we present a structural 01X-SAT-solver that combines the above mentioned encoding approach with a semantical node selection heuristics. Experimental results show that using this heuristics clearly outperforms the previous approaches both in CPU time used and in number of aborted instances.

Furthermore, we give for the first time a concise first-order formulation for the bounded model checking problem for blackbox designs with combinational blackboxes. To get correct results, the input-output-behavior of the blackboxes for equivalent input assignments within different time frames must be taken into account. This constraint leads to complex QBF formulas and we will present preliminary experimental data from the QBF Evaluation 2006 [6] showing that it requires sophisticated QBF-techniques to solve the corresponding QBF formulas.

The paper is structured as follows. In the following section, we review related work. In Section 3 we briefly give preliminaries necessary to understand the subsequent sections. Then, in Section 4 we describe our improved node selection heuristics when using 01X-logic to analyze blackbox problems. In this section we also present experimental results for our proposed optimization. Afterwards, in Section 5 we describe in detail how our QBF formulation works and present experimental results from the QBF solver evaluation in 2006. Finally, Section 6 concludes the paper.

---

*This work was partly supported by the German Research Council (DFG) as part of the Transregional Collaborative Research Center "Automatic Verification and Analysis of Complex Systems" (SFB/TR 14 AVACS). See www.avacs.org for more information.

2 Related Work

Model Checking of Blackbox Designs A first attempt for combinational equivalence checking of incomplete designs was made in [7] and further extended in [8]. In the context of symbolic CTL model checking, it turned out in [4] that modelling blackboxes by non-deterministic inputs ends up in ambiguous results when using different symbolic model checkers, e.g. VIS and SMV. Hence, in [4] approximation techniques were developed that give consistent results w.r.t. validity and realizability of incomplete designs. As already mentioned, in [5] a *bounded* model checking approach was presented that adapted the concept of Z-simulation, as it was used in [4] for BDD-based symbolic model checking, to the SAT-based bounded model checking framework. Recently, in [9] counterexample extraction within a BDD-based blackbox model checking framework was considered.

01X-Logic 01X-logic has one of its origins in ATPG where it was applied in circuit simulation to model the unpredictability of latches upon memory initialisation [10]. In [2] a transformation scheme was presented that compiles problems described in 01X-logic into pure propositional problems, allowing to use arbitrary engines that decide propositional logic.

Quantified Boolean Formulas Motivated by the impressive improvements for SAT engines within the last decade [11, 12, 13], current research tries to transfer this knowledge to the PSPACE-complete problem of deciding the validity of QBF [6]. Although current QBF solvers cannot compete generally with dedicated solvers for individual problems, the efficiency of modern QBF solvers has improved over the last years. It is also one aim of our work to trigger QBF-related issues and to support the QBF research community. The quest for the best algorithmic principle to decide QBF formulas is still open. Current QBF solvers are search-based in a DPLL-style (e.g., [14] and [15], expansion-based (e.g., [16]), or based on skolemization (e.g., [17]). Additionally, the usage of preprocessing techniques (see [18]) and the transformation into non-prenex form (see [19]) seem to be mandatory techniques to solve real world problems.

3 Preliminaries

We consider the analysis of sequential circuits where parts of the combinational logic are unknown, i.e., the circuit contains one or more blackboxes. In Figure 1 this scenario is visualized.

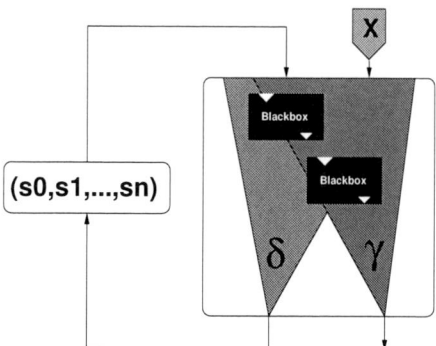

Figure 1. Sequential circuit containing combinational blackboxes.

3.1 Applications of Blackboxes

The concept of blackboxes can be used for various applications within a circuit design flow:

1. *Early Design Stage Verification.* Assume that two or more people are involved in designing a new circuit. Verification of properties or finding bugs, respectively, already in an early stage of the design, where at least one person has not finished yet its corresponding module, can be done by *blackboxing* the unimplemented module. Then, when a bug can be found independent of the blackbox, there must be an error outside the blackbox. Finding bugs in an early design stage can considerably save development costs.

2. *Module Abstraction.* Assume that a circuit developer wants to check some design property that does not depend on some modules of the design. These modules can be *blackboxed*, resulting in an abstraction of the circuit. This technique can be used, for example, when abstracting complex modules like multipliers or memory, decrease the required computational resources and thus enable the verification.

3. *Error Diagnosis.* Assume a bug was found and the circuit designer wants to know in which part of the design the error may be located. To do so, the designer can *blackbox* some candidate region in which the error is suspected. When blackbox analysis techniques can show that no error is observable, then it can be concluded that the candidate region contains the error location.

Modelling of the blackbox behaviour can be done in several ways. The most simple scheme relies on 01X-logic, a more advanced and more expressive scheme to be presented in Section 5 relies on Quantified Boolean Formulas.

$\text{AND}_{01X}(a,b)$				$\text{OR}_{01X}(a,b)$				$\text{NOT}_{01X}(a)$	
a \ b	0	1	X	a \ b	0	1	X	a	
0	0	0	0	0	0	1	X	0	1
1	0	1	X	1	1	1	1	1	0
X	0	X	X	X	X	1	X	X	X

Table 1. Boolean operations AND, OR, **and** NOT **extended to 01X-logic.**

3.2 01X-Logic

01X-logic extends propositional logic by a third logical value, X, to provide means to talk about the uncertainty of the status of propositional variables. Basic functions like conjunction, disjunction, and negation can be adapted conservatively by taking the controlling values of the gate function into account, as it is shown in Table 1.

The transformation scheme of [2] maps the three logical values $0, 1$, and X to binary tuples $(1,0), (0,1)$, and $(0,0)$, respectively. Using this mapping, the extended operators can be adapted to this encoding:

$$\text{AND}_{0IX}((g_0, g_1), (f_0, f_1)) := (g_0 + f_0, g_1 \cdot f_1)$$
$$\text{OR}_{0IX}((g_0, g_1), (f_0, f_1)) := (g_0 \cdot f_0, g_1 + f_1)$$
$$\text{NOT}_{0IX}((g_0, g_1)) := (g_1, g_0)$$

3.3 Quantified Boolean Formulas

Quantified Boolean Formulas (QBFs) extend propositional formulas by allowing to *existentially* ($\exists$) or *universally* ($\forall$) quantify some variables. A QBF of the form $\exists x : f(x)$ ($\forall x : f(x)$) is called *valid* iff for some (all) value(s) of $x \in \{0,1\}$, $f(x)$ evaluates to 1. Briefly, a QBF has the following general form:

$$Q_1 X_1 Q_2 Y_1 Q_3 X_2 Q_4 Y_2 \ldots Q_n Y_{\frac{n}{2}} Q_{n+1} X_{\lceil \frac{n+1}{2} \rceil} :$$
$$f(X_1, Y_1, X_2, Y_2, \ldots, Y_{\frac{n}{2}}, X_{\lceil \frac{n+1}{2} \rceil}),$$

where $X_1, Y_1, \ldots, Y_{\frac{n}{2}}, X_{\lceil \frac{n+1}{2} \rceil}$ are pairwise disjoint sets of propositional variables, $Q_1 = \exists, Q_i \in \{\exists, \forall\}, Q_i \neq Q_{i+1}, 1 \leq i \leq n$, and f is a propositional formula in conjunctive normal form (CNF). Wlog. we require the first quantifier to be existential, but X_1 may be empty, and the last quantifier to be existential, since innermost universally quantified variables can be immediately eliminated when f is given in conjunctive normal form. This form is typically used as input format by current QBF solvers.

3.4 And-Inverter-Graphs

The usage of And-Inverter-Graphs (AIGs) was mainly triggered by the work of Kuehlmann et al. [20] for space-efficient representation of boolean functions in the context

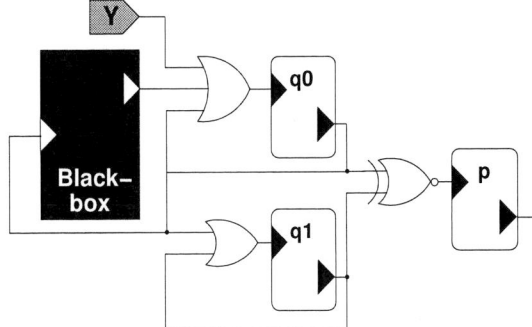

Figure 2. Example showing feasibility of 01X-logic to detect counterexamples.

of combinational equivalence checking and property checking. AIGs can be seen as combinational circuits restricted to two-input AND-gates which correspond to the vertices of the graph, and INV-gates which correspond to marked edges in the graph. The leaves of the graph are primary inputs. There exist efficient synthesis techniques based on structural and functional hashing to compute manageable AIGs for large circuits [20]. Figure 3 shows on the right two AIGs for the encoding of AND_{0IX}.

4 Improvements to 01X-based BMC

We start by looking at a small example showing that counterexamples can be found using 01X-logic. In Figure 2, a small sequential circuit with three flip-flops and one blackbox is shown. The next state functions, using 01X-logic, can be written as follows:

$$
\begin{aligned}
q_0' &:= q_0 + Y + X \\
q_1' &:= q_0 + q_1 \\
q_2' &:= \neg(q_0 \oplus q_1)
\end{aligned}
\tag{1}
$$

The property that should be checked is $\text{AG}(q_0 \oplus q_1) = \text{AG}(\neg p)$, i.e., p should always be 0. We simulate the circuit under the semantics of 01X-logic, i.e., in each time step we assign the logical value X to the output of the blackbox. Starting in the initial state ($q_0 = 0, q_1 = 1, p = 0$), the following table depicts a trace of length 3 that ends up in a state where $p = 1$ holds, so that the property is violated:

step	y	q_0	q_1	p
0	—	0	1	0
1	1	1	1	0
2	0	1	1	1

In [5] the first author has implemented a SAT-based framework for bounded model checking using AIGs and a structural SAT-solver using AIGs as basic representation.

1550-4093/07 $25.00 © 2007 IEEE

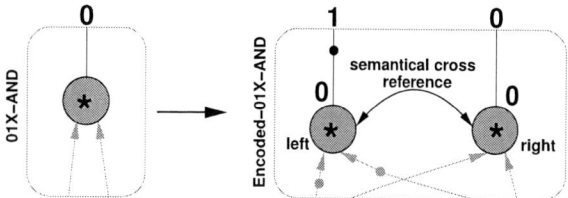

Figure 3. Semantical cross reference between AIG vertices.

The experiments in that work revealed that the encoding approach of Jain et al. [2] in general performs much more efficient than lifting the deduction rules of the structural SAT-solver to 01X-logic. But the experiments also showed that in some cases the SAT-solver can be misguided when using the encoding approach, because the variable selection mechanism of the propositional SAT-solver is not aware that the problem was generated from a 01X-logic problem. Put another way, the knowledge that two AIG vertices correspond to each other due to the binary encoding, was not available. Hence, one contribution of this work is to show that it pays off to take this correspondence into account during the search of the SAT-solver. To overcome the limitations of the previous approach, we propose an improved node selection heuristics that is used within our structural SAT-solver as follows.

The basic principle for our node selection heuristics is depicted in Figure 3. During construction of the AIG for the unfolding of the circuit under analysis, the two AIG vertices that are generated due the encoding approach are linked together by a semantical cross reference. The main drawback of the "blind" SAT-solver in [5] is that justifications of those two AIG vertices can be resolved at very different time points during the search. But now the semantical cross reference can be used to immediately handle AIG vertex justifications whenever the cross-referenced one was resolved. As an example have a look to the AIGs that refer to the encoding of an AND-gate, as depicted on the right hand side in Figure 3. I.e., assume that the justification of the left node was already resolved and recall that from the perspective of 01X-logic, we would like to justify the $0_{01X} = (1,0)$ value(s) of the encoding. Due to the semantical cross reference, we now can directly detect that the justification of the right AIG vertex should be handled. This strategy was integrated into the maximum-decision-level-heuristics already used in [5].

Table 2 shows experimental results for our new node selection heuristics.[1] As can be seen, it clearly outperforms the previous 01X-SAT-solver [5] not only in CPU time but also in the number of instances that can be solved.

---

[1]The CPU time limit was set to 900seconds. The times in column "Total" include the time for the aborted instances.

Solver	Time		#Solved / #Total
	Solved	Total	
"blind" SAT-solver [5]	964	17165	2712 / 2730
improved SAT-solver	386	12084	2717 / 2730

Table 2. Results for the semantical node selection heuristics.

5 QBF formulation

Before going into details how our QBF formulation works, we'll have a look at a small example, showing that with 01X-logic, it is not always possible to detect a counterexample.

Inaccuracy of 01X-Logic For the circuit in Figure 4 the next state functions are as follows wrt. 01X-logic:

$$
\begin{aligned}
q_0' &:= q_0 + X \\
q_1' &:= q_0 \cdot X \\
q_2' &:= 1 \\
q_3' &:= q_2 \\
p' &:= (q_0 + (\neg q_1)) \cdot q_3 \cdot y
\end{aligned}
\tag{2}
$$

The circuit is analyzed with respect to the property $AG(y \cdot q_3 \cdot (\neg q_1 + q_0)) = AG(\neg p)$, i.e., p should always be 0. The following trace of length 4 shows that starting from the state $(q_0 = 0, q_1 = 0, q_2 = 0, q_3 = 0, p = 0)$, in the last step the signal value of $(\neg q_1 + q_0)$, that is necessary to be 1 so that p can be forced to 1, evaluates to X, and hence the property cannot be falsified.

step	y	q_0	q_1	q_2	q_3	p
0	—	0	0	0	0	0
1	—	X	0	1	0	0
2	—	X	X	1	1	0
3	1	X	X	1	1	X

As we will see, using QBF it is possible to detect a counterexample for the example given in Figure 4.

Blackbox handling for QBF Besides the issue of over-approximation as sketched in the example, the 01X-logic approach also abstracts from the fact that different blackbox outputs may compute different boolean functions. To take this into account, disjoint variables are introduced for the blackbox inputs and outputs. I.e., for the ith blackbox $\mathcal{B}_i$ that has k inputs (l outputs), we introduce variables $(\xi_i^1, \xi_i^1, \ldots, \xi_i^k)$ for the inputs, and variables $(\gamma_i^1, \gamma_i^1, \ldots, \gamma_i^l)$ for the outputs. We abbreviate these vectors using Ξ_i and Γ_i, respectively. Since we look at a finite unfolding, we additionally attach a *time index* to the variables, i.e., the kth input of blackbox $\mathcal{B}_i$ in time frame t corresponds to the variable $\xi^k_{(i,t)}$. The notion for the vectors is extended in the same manner, i.e., $\Gamma_{(i,t)}$ denotes the vector of variables that correspond to the outputs of blackbox $\mathcal{B}_i$ in time frame t. In the following, we use $in(\mathcal{B}_i)$ ($out(\mathcal{B}_i)$) to denote the number of inputs (outputs) of blackbox $\mathcal{B}_i$.

Transition Relation with Blackboxes   The above introduced variables for the blackbox outputs can now be taken into account when building the transition relation. I.e., the transition relation is built by symbolic simulation using AIGs as underlying data structure. During symbolic simulation, the blackbox outputs are handled as additional primary inputs. Regarding the implementation, we construct a *generic* transition relation $T^{\mathrm{gen}}(s,x,\Gamma_1,\ldots,\Gamma_\beta,s')$ using generic versions Γ^{gen}_i of variables for the blackbox outputs of blackbox $\mathcal{B}_i$. For the construction of the finite unfolding of the transition relation, we use a substitution operator to replace all generic variables by their timed instantiation. In more detail, the transition relation $T(s_i,x_i,\Gamma_{(1,i)},\ldots,\Gamma_{(\beta,i)},s_{i+1})$, describing the transitions from time frame i to time frame $(i+1)$, is obtained as follows:

$$
T(s_i,x_i,\Gamma_{(1,i)},\ldots,\Gamma_{(\beta,i)},s_{i+1}) :=
$$
$$
T^{\mathrm{gen}}(s,x,\Gamma_1,\ldots,\Gamma_\beta,s') \Big|
\begin{array}{l}
s_i \leftarrow s \\
\Gamma_{(1,i)} \leftarrow \Gamma^{\mathrm{gen}}_1 \\
\Gamma_{(2,i)} \leftarrow \Gamma^{\mathrm{gen}}_2 \\
\ldots \\
\Gamma_{(\beta,i)} \leftarrow \Gamma^{\mathrm{gen}}_\beta \\
s_{(i+1)} \leftarrow s'
\end{array}
\quad (3)
$$

We apply the concept of *shadow*-variables suggested in [21] to efficiently implement this substitution. Now, the finite unfolding $T(\beta,d)$ of the transition relation up to depth d, that takes β-many blackboxes into account, can be constructed:

$$
T(\beta,d) := \bigwedge_{i=1\ldots(d-1)} T(s_i,x_i,\Gamma_{(1,i)},\ldots,\Gamma_{(\beta,i)},s_{i+1}) \quad (4)
$$

Input-Output-Consistency   Since we focus on combinational blackboxes, we have to take care about the deterministic input-output-behaviour of the blackbox. Put another way, since the blackbox implements a function rather than a finite state machine, it has to produce the same output values given the same input values. Hence, we require a predicate to ensure this consistency constraint. For a blackbox $\mathcal{B}_i$ and two different time frames t_1 and t_2, $t_1 \neq t_2$, the following predicate $\mathrm{IOC}(\mathcal{B}_i,t_1,t_2)$ is true iff the consistency constraint is not violated:

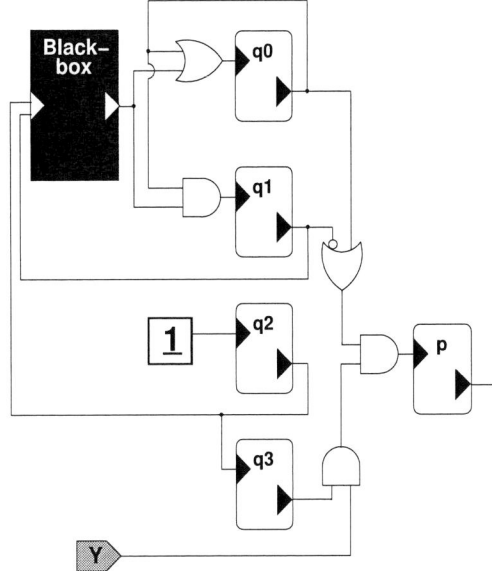

Figure 4. Example showing infeasibility of 01X-logic, but feasibility of QBF to detect counterexamples.

$$
\mathrm{IOC}(\mathcal{B}_i,t_1,t_2) := \qquad\qquad\qquad (5)
$$
$$
\left(\bigwedge_{m=1\ldots in(\mathcal{B}_i)} \xi^m_{(i,t_1)} \equiv \xi^m_{(i,t_2)} \right) \rightarrow \left(\bigwedge_{n=1\ldots out(\mathcal{B}_i)} \gamma^n_{(i,t_1)} \equiv \gamma^n_{(i,t_2)} \right)
$$

This consistency constraint has to be fulfilled for all possible time frame combinations for a given maximum unfolding depth d. We capture this constraint using the predicate $\mathrm{IOC}(\mathcal{B}_i,d)$:

$$
\mathrm{IOC}(\mathcal{B}_i,d) := \bigwedge_{\substack{1 \le t_1,t_2 \le d, \\ t_1 \neq t_2}} \mathrm{IOC}(\mathcal{B}_i,t_1,t_2) \quad (6)
$$

Finally, the requirement for consistent input-output-behaviour must be satisfied for all blackboxes. Let β be the number of blackboxes in the original sequential circuit. Then, $\mathrm{IOC}(\beta,d)$ is true iff the consistency constraint for all blackboxes holds in the finite unfolding up to depth d:

$$
\mathrm{IOC}(\beta,d) := \bigwedge_{b=1\ldots\beta} \mathrm{IOC}(\mathcal{B}_b,d) \quad (7)
$$

QBF Formula   Now we describe a QBF formulation for detecting counterexamples. Our QBF formula is *simulation driven*, since it follows the temporal behaviour of the sequential circuit and its finite unfolding, respectively. Put another way, we start in an initial state and select some assignment for the primary inputs. The values of the primary input and state variables induce deterministically values at the blackbox inputs. The QBF-solver then has to check that *for all* possible value combinations at the blackbox outputs a next state is reached from which, again, a sequence of input assignments can be found such that a state violating the property is reached. For different instantiations of the same blackbox, but in different time frames, the input-output-consistency described above must be ensured, to avoid *false negative* counterexamples. Finally, we are able to construct a formula $\varphi_{BMC}(d)$:

$$
\begin{aligned}
\varphi_{BMC}(d) := \\
\exists s^1 \exists x^1 \exists \Xi_{(1,1)} \forall \Gamma_{(1,1)} \ldots \exists \Xi_{(\beta,1)} \forall \Gamma_{(\beta,1)} \\
\exists s^2 \exists x^2 \exists \Xi_{(1,2)} \forall \Gamma_{(1,2)} \ldots \exists \Xi_{(\beta,2)} \forall \Gamma_{(\beta,2)} \\
\ldots \\
\exists s^{d-1} \exists x^{d-1} \exists \Xi_{(1,d-1)} \forall \Gamma_{(1,d-1)} \ldots \exists \Xi_{(\beta,d-1)} \forall \Gamma_{(\beta,d-1)} \\
\exists s^d : \\
IOC(\beta,d) \rightarrow \Big(I(s^1) \cdot T(\beta,d) \cdot (\neg P(s^d)) \Big)
\end{aligned}
\tag{8}
$$

$I(s^1)$ is a predicate for the initial states, and $P(s^d)$ describes the safety property in time frame d, i.e., $\neg P(s^d)$ states that the state reached after d time steps violates the given property. When formula $\varphi_{BMC}(d)$ is *true*, then there exists a counterexample (in terms of a Q-model, see [15]) of length d such that for every blackbox implementation there exists an individual counterexample that corresponds to a path in the Q-model of $\varphi_{BMC}(d)$.

Example Revisited   Let's see how the example from Figure 4 is handled with a QBF-solver using our encoding described above. Since within our QBF formulation, the blackbox outputs are handled by individual variables, the next state functions are rewritten, using the variable γ for the blackbox output.

$$
\begin{aligned}
q_0' &:= q_0 + \gamma \\
q_1' &:= q_0 \cdot \gamma \\
q_2' &:= 1 \\
q_3' &:= q_2 \\
p' &:= (q_0 + (\neg q_1)) \cdot q_3 \cdot y
\end{aligned}
\tag{9}
$$

Figure 5 depicts a decision tree that is implicitly built by a QBF-solver when analyzing the QBF formula $\varphi_{BMC}(3)$ for this example.

The left (right) edges correspond to assigning the blackbox output at some time frame to value 0 (1). The time frame can be derived from the depth of the source node of the edge. In the example, only in the last step the value of the primary input is of interest, hence in the diagram it is depicted on the bottom level only.

Solver	Time	#Solved/#Total
2clsQ	16828	0 / 28
GRL	16220	1 / 28
openQbf	16826	0 / 28
preQuantor	571	0 / 28
Qbfl	16792	0 / 28
Quaffle	16380	0 / 28
QUANTOR	906	0 / 28
QUANTOR_hc	900	0 / 28
qube3.0	16216	1 / 28
qube4.0	15828	1 / 28
qube5.0	20	28 / 28
semprop	16229	1 / 28
sKizzo-0.9-abs	9183	0 / 28
sKizzo-0.9-grn	2191	0 / 28
sKizzo-0.9.std	10761	0 / 28
SQBF	11359	0 / 28
sSolve	16808	0 / 28
ssolve+ut	16809	0 / 28
ssolve-ut	16809	0 / 28
WalkQSAT	16227	1 / 28
yQuaffle	16699	0 / 28

Table 3. Short track results from QBFEVAL'06 for the `blackbox_design` family [6, 22].

The decision tree can be read as follows: All path starting in the initial state lead to a state within 3 time steps whereby (1) the signal values along this path satisfy the input-output-consistency of Equation (7), and (2) the reached state violates the analyzed property, i.e., $p = 1$. Hence, the sequence $(-, -, 1)$ of values assigned to the primary input y is a counterexample. Please note that in general the counterexamples detected by our QBF formulation has not be uniform as it is the case for this small example.

Results from QBF Evaluation 2006   Table 3 shows the short track results from the QBF Evaluation 2006 [6]. The benchmarks used are derived from our problem setting and are denoted `blackbox_design`. The benchmark set consists of 28 instances describing a blackbox bounded model checking problem for the `PicoJava/biu` benchmark from the VIS benchmark suite. The QBF instances are available from [22]. As can be seen from Table 3, only one out of 21 QBF-solvers, namely qube5.0 (see [19] for current development of the QBF-solver qube), is able to solve all of the QBF instances. qube5.0 applies transformations into a non-clausal representation and applies simplification and rewriting of the quantifier tree resulting in a non-prenex QBF. Additionally, qube5.0 does efficient preprocessing of the original QBF, which drastically reduces the complexity of the QBF.

1550-4093/07 $25.00 © 2007 IEEE

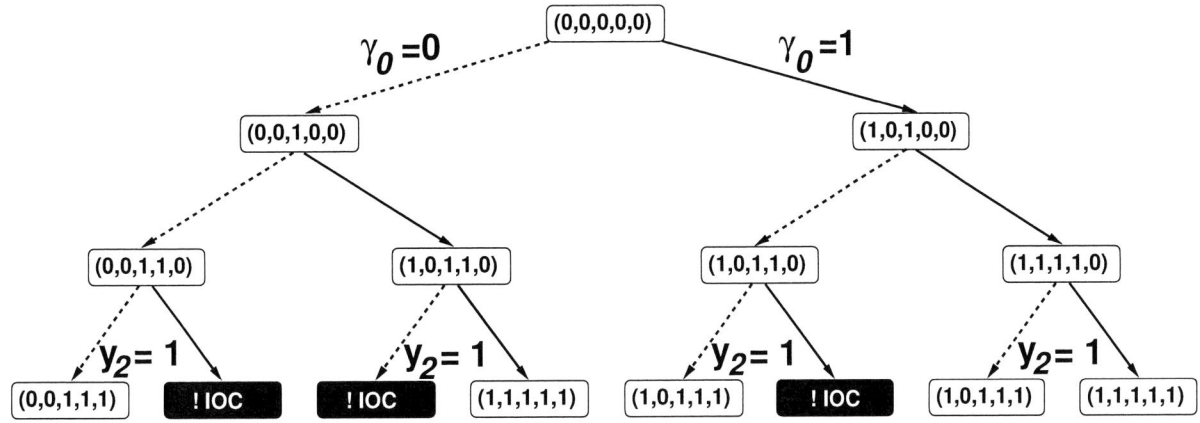

Figure 5. Decision tree implicitly built by QBF-solver. Tuples denote (q_0, q_1, q_2, q_3, p) **and the depth of a node corresponds to the time frame within the unfolding.**

6 Conclusions

In this paper we have reported on (1) optimizations for 01X-based bounded model checking of blackbox designs that considerably increase the efficiency of our 01X-SAT-solver, and (2) on a formalization of counterexamples using quantified boolean formulas that is more expressive than using 01X-logic. The QBF formulas turned out to be hard-to-solve for current state-of-the-art QBF solvers, but especially preprocessing techniques seem to make such a QBF-based approach viable.

As future work, we will investigate on how to combine 01X-logic and QBF. Furthermore, we will have a look on how the QBF formalization can be generalized to trade off the accuracy of the counterexamples and computational resources required to decide the existence of such a counterexample.

Acknowledgements. We are deeply grateful to Massimo Narizzano, Luca Pulina and Armando Tacchella for providing us the short track results of the QBF Evaluation 2006. Additionally, we would like to thank Tobias Nopper for contributing to the examples used in this paper.

References

[1] E. M. Clarke, A. Biere, R. Raimi, and Y. Zhu, "Bounded model checking using satisfiability solving," *Formal Methods in System Design*, vol. 19, no. 1, pp. 7–34, 2001.

[2] A. Jain, V. Boppana, R. Mukherjee, J. Jain, M. Fujita, and M. Hsiao, "Testing, Verification, and Diagnosis in the Presence of Unknowns," in *Proc. of VLSI Test Symposium*, 2000, pp. 263–269.

[3] C. Scholl and B. Becker, "Checking equivalence for partial implementations," in *Proc. of Design Automation Conference (DAC)*, 2001, pp. 238–243.

[4] T. Nopper and C. Scholl, "Approximate symbolic model checking for incomplete designs," in *Proc. of 5th International Conference on Formal Methods in Computer-Aided Design (FMCAD)*, Nov 2004, pp. 290–305.

[5] M. Herbstritt and B. Becker, "On SAT-based Bounded Invariant Checking of Blackbox Designs," in *Proc. of Microprocessor Test and Verification Workshop (MTV)*. Austin (TX), USA: IEEE Computer Society, 2005, pp. 23–28.

[6] M. Narizzano, L. Pulina, and A. Tacchella, "QBF Evaluation 2006," available on-line at www.qbflib.org/qbfeval [2006-08-02].

[7] W. Günther, N. Drechsler, R. Drechsler, and B. Becker, "Verification of designs containing black boxes," in *EUROMICRO*, 2000, pp. 100–105.

[8] C. Scholl and B. Becker, "Checking equivalence for circuits containing incompletely specified boxes." in *Proc. of 20th International Conference on Computer Design (ICCD)*, Freiburg im Breisgau, Germany, 2002, pp. 56–63.

[9] T. Nopper and C. Scholl, "Counterexample generation for incomplete designs," in *ITG/GI/GMM-Workshop "Methoden und Beschreibungssprachen zur Modellierung und Verifikation von Schaltungen und Systemen"*, 2007, to appear.

1550-4093/07 $25.00 © 2007 IEEE

[10] M. Abramovici, M. Breuer, and A. Friedman, *Digital Systems Testing and Testable Design*. Computer Science Press, 1990.

[11] J. Marques-Silva and K. Sakallah, "GRASP: A search algorithm for propositional satisfiability," *IEEE Trans. on Comp.*, vol. 48, no. 5, pp. 506–521, 1999.

[12] M. Moskewicz, C. Madigan, Y. Zhao, L. Zhang, and S. Malik, "Chaff: Engeneering an efficient SAT solver," in *Proc. of Design Automation Conference (DAC)*, 2001.

[13] N. Eén and N. Sörensson, "An extensible sat-solver." in *Proc. of 6th International Conference on Theory and Applications of Satisfiability Testing (SAT)*, ser. Lecture Notes in Computer Science, vol. 2919. Springer, 2004, pp. 502–518, selected Revised Papers.

[14] E. Giunchiglia, M. Narizzano, and A. Tacchella, "Qube++: An efficient qbf solver." in *Proc. of 5th International Conference on Formal Methods in Computer-Aided Design (FMCAD)*, ser. Lecture Notes in Computer Science, vol. 3312. Austin, Texas, USA: Springer, 2004, pp. 201–213.

[15] H. Samulowitz and F. Bacchus, "Using sat in qbf." in *Proc. of 11th International Conference on Principles and Practice of Constraint Programming (CP)*, ser. Lecture Notes in Computer Science, vol. 3709. Sitges, Spain: Springer, 2005, pp. 578–592.

[16] A. Biere, "Resolve and expand." in *Proc. of 7th International Conference on Theory and Applications of Satisfiability Testing (SAT)*, ser. Lecture Notes in Computer Science, vol. 3542. Vancouver, BC, Canada: Springer, 2005, pp. 59–70, selected Papers.

[17] M. Benedetti, "skizzo: A suite to evaluate and certify qbfs." in *Proc. of 20th International Conference on Automated Deduction (CADE)*, ser. Lecture Notes in Computer Science, vol. 3632. Tallinn, Estonia: Springer, 2005, pp. 369–376.

[18] H. Samulowitz, J. Davies, and F. Bacchus, "Preprocessing qbf." in *Proc. of 12th International Conference on Principles and Practice of Constraint Programming (CP)*, ser. Lecture Notes in Computer Science, vol. 4204. Springer, 2006, pp. 514–529.

[19] E. Giunchiglia, M. Narizzano, and A. Tacchella, "Quantifier structure in search based procedures for qbfs." in *Proc. of Conference on Design, Automation and Test in Europe (DATE)*. Munich, Germany: European Design and Automation Association, 2006, pp. 812–817.

[20] A. Kuehlmann, V. Paruthi, F. Krohm, and M. M.K. Ganai, "Robust Boolean Reasoning for Equivalence Checking and Functional Property Verification," *IEEE Trans. on CAD*, 2002.

[21] A. Kuehlmann, "Dynamic transition relation simplification for bounded property checking." in *Proc. of International Conference on Computer-Aided Design (ICCAD)*. San Jose, CA, USA: IEEE Computer Society / ACM, 2004, pp. 50–57.

[22] M. Herbstritt, "QBF `blackbox_design` family benchmarks," available on-line at http://www.qbflib.org/suite_detail.php?suiteId=22 [2006-08-02].

Embedded Software Validation: Applying Formal Techniques for Coverage and Test Generation

Tamarah Arons, Elad Elster, Terry Murphy, and Eli Singerman

Intel Corporation, email: first.second@intel.com

Abstract— the validation of embedded software in VLSI designs is becoming increasingly important with their growing prevalence and complexity. In this paper we present a new, hybrid, automated, validation methodology combining formal techniques and simulation. We introduce compositional approach to generate a formal model of the design, and show how the list of its feasible paths can be extracted. This list is then used for coverage metrics, and for test generation. This method has been successfully applied to complex microcode of a state-of-the-art microprocessor, and it is applicable to other classes of embedded software. Its effectiveness and scalability was demonstrated on a set of complex IA32 instructions, where unknown bugs have been detected and validation convergence time was reduced from weeks in a previous project to a matter of days.

Index Terms— Software verification and validation, Test generation, Formal methods

I. INTRODUCTION

Advances in software and hardware technologies make it possible to design and manufacture products of increasing complexity. New design technologies often necessitate new validation techniques, which must provide a high degree of confidence in the system correctness. This is especially true for complex software systems embedded in safety critical applications where failures can cause fatalities and for complex hardware systems where bugs discovered too late carry a hefty price tag.

Formal verification (FV), which gives full coverage of the verified system, has been attracting considerable attention. Over the past twenty years there has been significant progress in FV theory and algorithms both in academia and in industry. Today automatic FV tools are being applied as a common validation technique in both industrial strength software **Error! Reference source not found.** and hardware **Error! Reference source not found.**. However, despite their success in various domains, capacity limitations restrict the use of

automatic FV tools on very large models. Thus, the main validation vehicle both for hardware and for software remains simulation.

Simulation has the advantage of significantly better capacity, and thus provides the ability to handle large designs. However, coverage achieved by simulation of large systems is very limited and thus the major questions that have to be addressed in order to make simulation more effective are which behaviors needs to be covered and how to cover them.

Hybrid verification flows which combine FV and simulation are an attempt to benefit from the advantages of both worlds, providing improved validation of complex designs. In the bug-hunting hybrid techniques, offered by several EDA vendors, simulation traces are extended by FV exploration (like bounded model checking) around strategic locations in order to cover more design scenarios.

In this paper we describe a hybrid approach, exploiting formal methods to *improve traditional, simulation based validation techniques*. Our method, supported by fully-automatic tools, addresses two major problems in simulation based validation of large embedded software models: constructing relevant coverage targets and generating tests to hit them. While improvement in each of these dimensions is of high value, a combined approach as we propose and demonstrate in this paper provides even higher value.

The concrete problem domain we discuss in this paper is the validation of microcode of an industrial microprocessor. A complex microcode program can contain thousands of possible execution paths with hundreds of thousands of statements along the longer paths. Due to this complexity, deriving the set of possible paths (to serve as coverage goals) and generating tests to hit them are both complex tasks. We show how both problems can be addressed efficiently and automatically using formal techniques.

Although our method has been developed for microcode validation, we believe it will be applicable to other low-level embedded software. As platforms on a chip integrating multi-core CPUs and communication devices develop, we anticipate that they will include low-level software components controlling the interaction between the platform units. We see this as a growing area where techniques like those proposed in

1550-4093/07 $25.00 © 2007 IEEE

this paper are likely to be beneficial.

A. Overview of Microcode Validation

The concrete problem domain discussed in this paper is the validation of the microcode of an industrial microprocessor. Several state-of-the-art microprocessors have a CISC (Complex Instruction Set Computing) interface while their internal hardware implementation is that of RISC (Reduced Instruction Set Computing). To bridge this gap, these CPUs contain an embedded layer of *microcode*. This low-level embedded software translates between the architecture interface and the hardware implementation based on the micro-architecture. As CPUs become more complex so does their microcode. In a modern CPU, microcode programs may contain thousands of possible execution paths with hundreds of thousands of statements along the longer ones.

Two characteristics of microcode differentiate it from "typical" software. Firstly, microcode is very "machine aware", its execution being heavily dependent on changes in the microarchitectural state. Secondly, it is structurally modest, having simple data structures, no pointers, no recursion, less loops and is terminating. This structural simplicity makes it more amenable to validation methods often considered infeasible for general purpose high-level software.

There are many accepted coverage metrics for software. Path coverage is a structural (white box) testing strategy requiring that each of the execution paths in a program be tested. It is generally considered powerful but impractical as even small programs may have a huge (even infinite) number of paths **Error! Reference source not found.**. For microcode, since correctness is critical, path coverage has long been the chosen coverage metric. Generating the set of feasible paths, the coverage goal, is, however, a complex task requiring intensive human effort, as is generating tests to hit them. We have developed an efficient automated method for enumerating the path-space, and are developing a complementary automated method for test generation. While improvement in each of these domains is of high value, a joint approach, as we propose in this paper, provides even higher value.

Intuitively, we use formal methods to generate the set of feasible paths. Thereafter, we employ a test generation methodology combining elements from both counter-example based test generation and randomization.

B. Related Works

The ideas of using symbolic simulation for verification of embedded software and of combining symbolic simulation and test generation are well known. See for instance **Error! Reference source not found.** and **Error! Reference source not found.** respectively.

Relevant to our work is the FSM-based approach used to generate assembly test programs **Error! Reference source not found.Error! Reference source not found.Error! Reference source not found.Error! Reference source not found.** to verify hardware. A finite state machine describing the control logic is used to abstract the behavior of the design, and a set of transitions covering tours are generated **Error! Reference source not found.**. These tours can then be concretized into a list of instruction types and used to generate test programs **Error! Reference source not found.**. Generating the FSM, either by annotation of the HDL model **Error! Reference source not found.**, or by experts writing it manually **Error! Reference source not found.Error! Reference source not found.Error! Reference source not found.Error! Reference source not found.**, is however a non-trivial exercise. To minimize the "expensive expert time" involved in the FSM generation, **Error! Reference source not found.** suggests using a set of communicating FSMs, and devoting considerable effort to "modeling guidelines" which will reduce the FSM size.

We also adopt methods from counter-example based test generation. In this paradigm, model checkers, possibly bounded, provide counter-examples from models of negations of properties. The witness of a counter-example is an initial test state, and can be used to generate tests **Error! Reference source not found.Error! Reference source not found.**.

Complex designs that combine hardware and embedded software have been the target of multiple test generation efforts. One ongoing program uses constraint-solving to generate architectural tests for PowerPC and IA32. The architecture is modeled by experts, and the modeling code is used as the basis for directed-random test generation **Error! Reference source not found.**. Other works present how coverage directed test generation can improve the quality of tests considerably **Error! Reference source not found.**, but this implies an existing coverage space, usually manually specified. Compared with these works, our testing strategy does not require the modeling of the entire architecture and does not require an existing coverage space.

The contribution of our approach is that it automatically handles significantly larger models (tens of thousands of transitions, versus hundreds of transitions). One reason for this is that we introduce a compositional method of generating a formal model. This reduces the manual effort required (compared to writing an FSM, for example) and allows additional programs in the same language to be verified with little additional effort. Symbolic simulation also brings significant capacity advantages. As opposed to FSMs and BDDs, symbolic simulation allows the simulation of control logic while preserving exact data values. The precise control determination is maintained without a need to enumerate all the possible data values.

The paper is organized as follows: We first introduce our methodology (Section 2), and then discuss its practical application to microcode validation (Section 3). In Sections 4 and 5 we report and evaluate results obtained thus far and discuss the advantages and disadvantages of our approach.

1550-4093/07 $25.00 © 2007 IEEE

II. METHODOLOGY

The methodology described in this work enables the construction of a path coverage space and the generation of directed tests for complex designs with minimal intervention. In this section we describe our methodology. Although the examples are drawn from microcode validation, the methodology presented is application independent. In the next section we describe the application of this methodology in microcode validation tools.

The method we present is four-phase: We first translate the source code into a formal model (formalization), and then simulate it symbolically. The feasible paths are extracted, and finally tests can be generated. We now describe each of these four stages in detail.

A. Building a Formal Model

The embedded system is written in a non-formal low-level software language. In addition to its explicit semantics, each instruction may have various side effects. For example, a memory read explicitly updates the target register with the memory contents. In addition, it implicitly checks the address and other paging-related information for faults.

Accurate symbolic simulation requires a formal model of the program in which all side-effects are made explicit. We propose a *compositional* approach in which every instruction in the source code has its own formal model, independent of its context (the program to be simulated). The formal model of an instruction is a list of simple, unambiguous operations, in an intermediate representation language, *IRL*. All side effects are modeled explicitly in this language **Error! Reference source not found.**.

The full IRL program is built by replacing each instruction in the source by its formal IRL model. This is the input to the symbolic simulation stage.

B. Symbolic Simulation

The IRL program is formally simulated so that all its feasible paths can be calculated and extracted. The system starts at a symbolic initial state, where each variable has a symbolic initial value. A constraint mechanism allows the user to constrain these initial values so as to start from a legal initial state e.g. if variables are mutually dependent, this dependency is formalized and added as a constraint.

Variable values are represented during simulation as symbolic expressions over the set of initial values and constants that appear in the program. Let the initial values of EAX, EBX and CPL in Figure 1 be the symbolic values EAX_0, EBX_0 and CPL_0 respectively. After simulating instruction I2, the value of EBX_1 is an expression over EAX_0: $((EAX_0 > 7) ? 8 : (EAX_0 - 2))$.

---

toy_instruction:
I1: if (CPL > 0) fault;
I2: if (EAX > 7) then EBX := 8 else EBX := EAX - 2;
I3: if (EBX < 5) goto skip_mask;
I4: EAX := EAX & 0x000F;
I5: skip_mask: if EAX < EBX fault;

Figure 1: A Simple IRL program

If IRL programs were merely lists of assignments, this would suffice. However, they also include conditional jump and fault statements. Because the state is symbolic, the value of the condition cannot always be equivocally determined. For example, at statement I3 the program checks the condition $(EBX_1 < 5)$ and then either continues or jumps to skip_mask. Checking the value of $(EBX_1 < 5)$ is equivalent to evaluating whether $((EAX_0 > 7) ? 8 : (EAX_0 - 2)) < 5$. We can find assignments to EAX_0 where the condition is true (6), and where it is false (10). Symbolic simulation builds a set of *symbolic paths* representing control distinct (as opposed to data distinct) paths. Each symbolic path records the current symbolic variable values, the path condition, and the exit name (in our example either fault or normal termination).

For example, after statement I1 has been simulated, we have 2 symbolic paths, one with path condition $(CPL_0>0)$ which has reached a fault, and the second, with path condition $(CPL_0=0)$. Simulation of the second path continues, with the updating of the value of EBX at statement I2. At statement I3 the path is again split in two: one with path condition $(CPL_0=0$ & $EBX_1 < 5)$, and the second with path condition $(CPL_0=0$ & $EBX_1 \geq 5)$.

The path space can be viewed as a tree (**Error! Reference source not found.** 2). The *path condition,* the condition under which a particular path from a root to a leaf (final state) is taken, is the conjunction of the conditions along the path. The conditions are marked in bold in the diagram. Thus, in our example we have 5 paths with the following path conditions:

P0: $(CPL_0>0)$ → fault
P1: $CPL_0=0$ & $(EBX_1 < 5)$ & $(EAX_0 < EBX_1)$ → fault
P2: $CPL_0=0$ & $(EBX_1 < 5)$ & $(EAX_0 \geq EBX_1)$ → end
P3: $CPL_0=0$ & $(EBX_1 \geq 5)$ & $(EAX_1 < EBX_1)$ → fault
P4: $CPL_0=0$ & $(EBX_1 \geq 5)$ & $(EAX_1 \geq EBX_1)$ → end

Final path conditions are expressed as functions of the initial state. For example, the path condition for P2 will be:

P2: $CPL_0=0$ & $(((EAX_0 > 7) ? 8 : EAX_0 - 2) < 5)$ & $(EAX_0 \geq ((EAX_0 > 7) ? 8 : EAX_0 - 2))$ → end

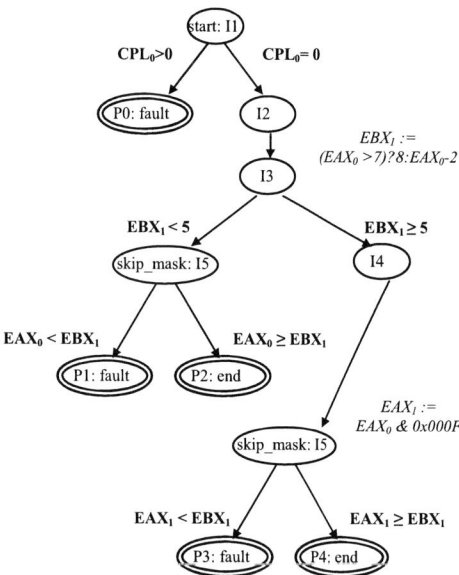

Figure 2: A Tree of Paths with Path Conditions

C. Path Extraction

Once the full simulation tree has been built, a SAT solver is used to check the feasibility of each path, by checking the satisfiability of its path condition. Path P1, for example, is found to be infeasible, because the path condition leading to it is unsatisfiable. For feasible paths, the SAT solver returns a witness, which corresponds to an initial state from which the path would be taken.

During simulation we also record the list of executed instructions and miscellaneous interesting information. This is then used to gather the requisite data for constructing the coverage space. For example, for path P2 we have the information: Instruction: I1 I2 I3 I5 end;

The list of executed source instructions is placed in a coverage database, against which simulated tests can be compared. The satisfying initial condition can be used to generate a test in which this path is taken.

D. Generating Tests from Initial States

The feasible execution paths are used as the coverage goals against which coverage is measured, thus addressing the first challenge in microcode validation. We now address the second challenge, generating assembly stimuli to hit the desired paths.

As mentioned above, we use a SAT solver to check the satisfiability of path conditions and thus determine the feasibility of paths. When the path condition is satisfiable, the solver returns a *witness* - a set of initial values under which the path condition is satisfied. This is, in fact, a set of values under which this specific path would be taken. For example, for path P2, we may get a witness of $CPL_0=0$, $EAX_0=6$. The witness can also be less specific, e.g. for P3 we could have

$CPL_0=0$, $EAX_0[3:2] = 0b11$. Various different values of EAX_0 (e.g. 12, 29) satisfy this requirement, and result in path P3 being taken.

These architectural constraints are the relevant values at the point in time at which the instruction was initiated. However, they do not represent the entire machine state, nor explain how to reach this state by a sequence of assembly instructions. We cannot simply "jam" these values into a simulator as they do not take into account relations that might have been the effect of previous instructions. Also, some testing environments, such as the system level, have low controllability and require architecturally-correct tests. Therefore, to actually exercise the path an architecturally-correct test must be created, in which the desired initial state is reached before the path is exercised. To bridge this gap between what state the uArch wants, and how to actually get there by executing assembly instructions, we apply global *test constraints* when running Ample. These are taken into account when Ample computes path feasibility. In addition, the witness Ample outputs for each path include assignments to the variables in these test constraints.

A given path in the embedded code will be tested many times during validation, and it is desirable that it be run in (hardware) environments that are as diverse as possible. The constraint expressions generated by the SAT solver typically specify only a small subset of the architectural state. The test generator uses randomization to ensure that each time a test for a specific path is generated; it is exercised from within a different context, thus covering more of the test space. For example, a variety of different values for EAX could be used, and the test could be run under various modes (such as real mode, protected mode and long 64-bit mode).

III. APPLYING THIS METHODOLOGY TO MICROCODE

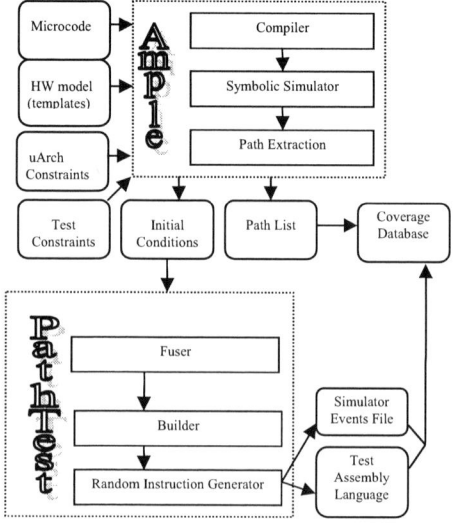

Figure 3: Ample - PathTest Flow Diagram

Intel CPUs implement the IA32 instruction set architecture (including various extensions, such as virtualization technology, VT). The IA32 architecture is CISC, meaning that a single instruction can implement a complex algorithm. Instructions are implemented in *microcode*, an assembly-like language which utilizes the underlying micro-architecture. Microcode programs ('microprograms') consist of discrete RISC micro-operations, *uops*. These uops are the connecting layer between architectural instruction behavior and the actual hardware.

Given the high cost of errors in embedded software, and IA32 microcode in particular, we try to exercise a large portion of the feasible microcode control paths. The objective of our work is to create a system that will take a microprogram as input, analyze it automatically, and create tests that will exercise the combined hardware/software design using all feasible control paths in this microprogram.

Two independent tools were developed: Ample and PathTest (

Figure 3). Ample (Automatic Microcode Path Logic Extraction) generates the list of microcode paths and initial states. PathTest uses these to generate microcode tests.

The input to Ample is the program source, and a set of *templates*, used for generating the formal model. Every template describes a specific uop and includes a formal representation, in IRL, of the operation done by that uop. The complexity of uops is such that each one is translated into multiple (even hundreds) of IRL statements. In addition, the user may provide a *constraints* file constraining the initial state to conform to the architecture. During *compilation*, the complete IRL program is generated from the microcode source and IRL templates.

In constructing a test PathTest takes into account both the initial state computed formally, and the architectural constraints associated with each element of the architecture. For example, in IA32 there are constraints ensuring that the memory regions dedicated to storing the test's page tables do not overlap with those used to store the test's object code (a test building restriction) and that paging be turned off when the CPU is not in protected mode. The fuser creates the constraint variables and expressions integral to the architecture and test, and combines them with the constraints produced by Ample. The solver solves these constraints, assigning values to constraint variables to satisfy each constraint. The builder uses these solved constraint variable values to create an actual architectural test – an IA32 test in our case.

A. Observations from the Application to Microcode

Generating the templates requires non-negligible effort, and in our first project took a number of person months. Given that the semantics of the design language do not change too significantly between projects, only those uops that are modified or added need to be updated between projects, and the effort becomes reasonable.

The correctness of the templates is crucial – template bugs introduce errors into the formal model, and may cause paths to be missed. To obviate this, we are developing a method for formally comparing the template modeling of each uop against its hardware specification. This uses a different formal technology, and is out of the scope of this paper.

Once the template basis has been written, new programs can be simulated using these templates. While some new programs can be simulated immediately, others do require user intervention (through additional directives in the constraints file). The primary sources of difficulties are loops, indirect jumps, and computational complexity.

Loops increase both the number of possible control paths and their complexity. Loops that are executed more than a specific number of times are flagged, and cause simulation to halt. The user must then aid the tool in dealing with the loop. Frequently, it is decided that it is sufficient for the loop to be executed a certain limited number of times, and an appropriate directive is given to the tool.

Indirect jumps are jumps to targets that are given by variable expressions e.g. goto EAX. Generally the tool succeeds in recognizing the potential target(s) in the expression, but when it does not the user is requested to provide this information.

Lastly, as with all formal methods, complexity sometimes becomes a problem. This is due to the sheer size of the model. Merge-points are an efficient mechanism for reducing complexity. These are points at which paths are merged together, transforming the path tree into a DAG; reducing the number of paths significantly. This is particularly useful when programs have independent "phases" e.g. parameter checking followed by execution. Identifying effective merge-points requires some understanding of the code, and currently must be done manually.

IV. RESULTS

We took a staged approach in developing and evaluating the technology. The path extraction and test generation components were developed and tested independently. When these components became functional the combined flow was evaluated.

A. Path Extraction Results

To evaluate the path extraction component (Ample) we chose several IA32 instructions of significant complexity. For these microprograms we had pre-existing path lists generated by a previous generation tool that used manual annotation of microprograms with control-path information to generate the list of feasible paths (*annotation approach*). We checked Ample's results against these existing lists of paths; each discrepancy was analyzed by a microcode validation expert.

The first step was to generate paths for these microprograms using Ample. Due to their complexity, some of the translated programs consisted of tens of thousands of IRL statements and several thousand paths. The effort required to generate these paths was small, requiring little or no manual intervention. Path generation time was several hours for the most complicated microprograms. Some examples are noted in

Table 1. Particularly noteworthy is that the time needed to generate one path remained relatively constant, even when the complexity of the instruction increased considerably.

We compared the results to the existing list of paths. Even though we had high confidence in the quality of the annotation-derived list of paths, Ample successfully detected that a relatively high number of paths were infeasible. More importantly, it revealed paths that were missing due to incorrect annotations.

To summarize these results, Ample was able to handle most microprograms with little intervention. We have so far applied Ample to 110 microprograms, 95 of which ran without requiring manual intervention. Some intervention was needed in the other 15 – generally additional constraints on the initial state, or loop abstraction. This is in contrast with our previous annotation approach which was time consuming and error prone. About 30% of the microcode validation effort used to be spent on writing and maintaining annotations. Moving to Ample-based coverage creation required a one-time effort to implement the IRL templates. From that point on, generating paths was mostly straightforward and the results were of much higher quality than using previous method. For the more complex instructions, simplifications (merge-points and loop abstraction) were required to reduce the exponential nature of the path coverage space. Given these simplifications, Ample proved capable of handling the most complex instructions of the IA32 ISA. For example, the effort of validating a complex microprogram was reduced from weeks in previous microprocessor to a matter of days using Ample in current microprocessor.

B. Ample/PathTest Results

Once the path extraction and test generation[1] components were both functional, we applied the combined product. We started with simple instructions to build confidence in the new technology. We then gradually moved our way up to the most complex instructions in the current microprocessor design. To date, we have used this method for generating tests for 60 microprograms. For each of these, assembly tests have been computed fully automatically, exercising all feasible execution paths. This resulted in the discovery of 10 previously unknown bugs. Encouraged by these results we are now applying this methodology for validating more instructions.

V. SUMMARY

In this paper, we described a two stage method for improving validation of embedded software using formal techniques. We first analyze the software program using

[1] Exercising and debugging PathTest stand-alone was done using dedicated test guiding annotations which were removed when we started applying the combined Ample/PathTest system.

formal techniques and generate a coverage space for all feasible control paths. Each control path is accompanied by initial state constraints sufficient to cause it to be taken. In the

Instruction	# of paths	# of statements in longest path	Generation time (sec.)	Generation time per path (sec)
InstrA	24	4606	489	20
InstrB	75	7430	1476	19
InstrC	79	8369	1392	18
InstrD	100	9622	7728	77
InstrE	102	13232	4397	43
InstrF	102	13233	5252	51
InstrG	407	14124	10707	26
InstrH	452	15552	28955	64
InstrI	538	33260	19722	36
InstrI + extra trap variables	8709	33260	73889	8.5

second stage these constraints are used to build architectural tests. The initial state constraints are combined with architectural constraints, and architecturally-correct tests are generated.

The contribution of our paper is twofold: we present a *compositional* modeling method – instead of manually modeling the high-level behavior of the system **Error! Reference source not found.**, we model the individual source instructions and the tool creates the high-level formal model. Furthermore, we show that our method is *scalable* and can be applied to very large designs. Modeling the micro-architecture is done only once and then used for the coverage construction of all microprograms that are based on the same micro-architecture. Other formal methods require user supplied descriptions for every program for which coverage construction is required. Thus, the modeling effort is fixed whereas in other methods it is linear in the number of programs.

Our approach is at a disadvantage in cases where the individual source instructions are very complex, such as floating-point and other numerical instructions. In these cases, coverage construction can be performed only when the needed modeling is in place. In contrast, the complexity involved in user supplied descriptions, e.g., FSMs or annotations, is fixed, and even programs that rely on complex micro-architectural properties can easily be annotated.

There are benefits and drawbacks to the type of tests generated by our system relative to those of other Automatic-Test-Generation (ATG) systems. A popular and effective method is the generation of random, user biased, tests **Error! Reference source not found.**. These tests are often long, with many permutations, and may reach interesting situations which were not targeted. The Ample/PathTest system aims its tests at specific control paths. While these tests will hit the desired targets, they may miss interesting untargeted scenarios. This is a classic black-box/white-box verification tradeoff. We recommend using randomized test generation to

Table 1: Ample Path Generation

collect initial coverage and reap the benefits of that testing approach. Once coverage stops climbing the white-box test generation methodology presented here becomes preferable, and can be used to cover hard-to-hit paths.

The methodology presented above was developed for, and applied to, IA32 microcode validation. We believe that it is generic and potentially applicable to other hardware designs that include embedded software. The symbolic simulation and path extraction phases run on IRL and are independent of microcode. While the compiler and the output format are tailored to our microcode application, Ample can be applied to other applications (e.g. firmware or device drivers) once the basic building blocks of the design are described in IRL

Future work in microcode domain includes allowing the user finer control over the coverage space creation and path generation (improved access to trap variables, for example). We are continuously working to increase the capacity of our system and are also looking at improving our test generation module by using random biasing to reach potentially hazardous situations more often.

We have recently started to work on adapting this technology to other types of embedded SW in Intel microprocessors.

ACKNOWLEDGEMENTS

We thank members of the Ample and PathTest development teams, past and present, for their invaluable contribution to this project.

REFERENCES

[1] A. Aharon, D. Goodman, M. Levinger, Y. Lichtenstein, Y. Malka, C. Metzger M. Molcho and G. Shurek. Test Program Generation for Functional Verification of PowerPC Processors in IBM, DAC'95

[2] T. Arons et al., Formal Verification of Backward Compatibility of Microcode, CAV'05, 2005.

[3] R. S. Boyer, B. Elspas and K. N. Levitt, SELECT - a formal system for testing and debugging programs by symbolic execution, 1975.

[4] T. Ball and S. K. Rajamani. Automatically Validating Temporal Safety Properties of Interfaces. *SPIN 2001*.

[5] F. Copty, L. Fix, R. Fraer, E. Guinchiglia, G. Kamhi, A. Tacchella, and M. Y. Vardi. Benefits of bounded model checking at an industrial setting. *CAV 2001*.

[6] D. W. Currie, A. J. Hu, S. Rajan, and M. Fujita, Automatic Formal Verification of DSP Software, 37th DAC, 2000.

[7] S. Fine and A. Ziv. Coverage Directed Test Generation for Functional Verification using Bayesian Networks, DAC'03

[8] D. Geist, M. Farkas, A. Landver, Y. Lichtenstein, S. Ur, and Y. Wolfsthal. Coverage directed test generation using symbolic techniques.,*FMCAD'96*, pages 143-158, 1996.

[9] A. Gupta, A. Casavant, P. Ashar and X. G. Liu. Property-Specific Testbench Generation for Guided Simulation, VLSID'02.

[10] G. Hamon, L. de Moura and J. Rushby, Generating efficient test sets with a model checker, *IEEE Conf. Software Eng. and Formal Methods*, pages 261-270, 2004.

[11] R. C. Ho, C. H. Yang, M. A. Horowitz, and D. L. Dill. "Architecture validation for processors", *Proc. Int. Symp. Computer Architecture (ISCA '95)*, pp. 404—413, 1995.

Instruction	# of paths	# of statements in longest path	Generation time (sec.)	Generation time per path (sec)
InstrA	24	4606	489	20
InstrB	75	7430	1476	19
InstrC	79	8369	1392	18
InstrD	100	9622	7728	77
InstrE	102	13232	4397	43
InstrF	102	13233	5252	51
InstrG	407	14124	10707	26
InstrH	452	15552	28955	64
InstrI	538	33260	19722	36
InstrI + extra trap variables	8709	33260	73889	8.5

[12] D. Lewin, D. Lorenz and S. Ur, A methodology for processor implementation verification, *FMCAD'96*, 1996.

[13] D. Lugato, F. Maraux, Y. Le Traon, V. Normand, H. Dubois, J. Y. Pierron and J. P. Gallois. Automated Functional Test Case Synthesis from THALES industrial Requirements, RTAS'04

[14] P. Mihsra and N. Dutt. Graph-Based Functional Test Program Generation for Pipelined Processors, DATE'04, pages 182-187, 2004.

[15] S. C. Nfatos. A Comparison of Some Structural Testing Strategies, IEEE Trans on Software Engineering 14(6), 1988.

[16] S. Ur and Y. Yadin, Micro Architecture coverage directed generation of test programs, *DAC'99*, 1999.

Challenges in System on Chip Verification

Noah Bamford, Rekha K Bangalore, Eric Chapman, Hector Chavez, Rajeev Dasari, Yinfang Lin,

Edgar Jimenez

Freescale Semiconductor Ltd, China, Mexico, USA

Abstract

The challenges of System on a Chip (SoC) Verification is becoming increasingly complex as submicron process technology shrinks die size, enabling system architects to include more functionality in a single chip solution. A functional defect refers to the feature sets, protocols or performance parameters not conforming to the specifications of the SoC. Some of the functional defects can be solved by software workarounds but some require revisions of silicon. The revision of silicon not only costs millions of dollars but also impacts time to market, quality, customer commitments. Working silicon for the first revision of the SoC requires a robust module, chip and system verification strategy to uncover the logical and timing defects before tapeout. Different techniques are needed at each level (module, chip and system) to complete verification. In addition Verification should quantify with a metric at every hierarchy to assess functional holes and address it. Verification metric can be a combination of code coverage, functional coverage, assertion coverage, protocol coverage, interface coverage and system coverage. A successful verification strategy also requires the test bench to be scalable, configurable, support reuse of functional tests, integration with tools and finally linkage to validation. The scope of this paper will discuss the verification strategy and pitfalls used in verification strategy and finally make recommendations for successful strategy.

I. INTRODUCTION

In 1965, Gordon Moore observed that number of transistors in a single integrated circuit was doubling every two years and he predicted that this exponential growth would continue in perpetuity. As advances in methodology, tools and process have enabled this growth, the task of verifying the functionality of the designs as also grown exponentially. The verification challenge are described below

- Not only is the number of transistors increasing, the functionality is becoming more complex as more features are added
- The cost of mistakes is increasing rapidly as the cost of masks increases
- Advances in computer aided design has enabled logic to grow at an exponential rate, which as resulted in the growth of the state space at exponential rate
- As State space for Verification is increasing, new tools and methodology are required

In addition to verifying the additional logic features added to the specification for each new generation of SoC, the market viability of the SoC must be insured. This means that its power characteristics must be acceptable and the performance must be adequate. It must be able to communicate with other chips and memories on a board and comply with a myriad of standards. Because of the increasing cost of tapeouts and the decreasing market window, all this must be verified in the pre-silicon design phase. In order to meet this challenge, verification methodology is continually improving. This paper discusses building a verification infrastructure to meet the verification challenge and how to insure the quality of SoC prior to tapeout.

II. FUNCTIONAL DEFECTS

Before examining how to design a verification infrastructure and insuring the quality of a SoC, it is helpful to examine the types of defects verification hopes to catch. Functional defects are deviations from the specification and can be classified as either "soft" OR "hard". This classification is based on the severity of the defect and how it is fixed. A defect is said to be soft if a feature is not being used or the software can be changed to work around the defect. After being fixed, the defect still exists in hardware but is no longer a problem for the end user. A hard defect is one which must be fixed by changing the implementation of the circuit. As a soft defect does not require a re-spin, the cost of fixing it is much lower.

An example of a soft defect is shown in the following case that was seen while verifying a baseband processor. In order to understand defect, a little background is necessary. The baseband includes a "Deep Sleep" mode where most peripherals clocked by high speed clock are shut down (this clock is turned off), and a low speed clock is used by those peripherals which need to remain awake.

The defect occurred in the module used to control the entrance and exit from this mode. During the exit of the sleep mode, the module stayed in its WARM state one clock cycle longer than expected. Because of this, other modules inside the SoC had sync-up problems resulting in a poor call quality. The original problem observed was that module was restarting one low speed clock cycle later than specified when exiting sleep. This caused the phone to resynchronize with the network. A software patch was proposed to workaround the problem and solves the synchronization issue. Because no hardware changes were required, the cost of the defect was not as great.

A hard functional defect occurs due to logic bugs, incorrect modeling, incorrect timing constraints, and asynchronous interfaces in the SoC or due to unexpected process variations found during manufacturing. A hard functional defect usually requires a re-spin of the SoC and results in customer quality and product delays.

An example of hard functional defects can occur with simple scenarios like power up sequence of the part. Analog modules used in SoC verification should be verified standalone with stand alone and with design marginalities for handling process variations. A power on reset an analog module cannot be represented with the spice model during regressions for the full SoC. The reset sequence of the SoC with a semi-accurate module can lead to unexpected scenarios in simulation. This includes timing of the chip coming out of reset and the start of code execution. A design fix was recommended to fix this problem.. It is also recommend to ensure debug modes are built into the part to handle such scenarios. This includes alternate functional modes that can be used by customer and product debug team.

III. ADDRESSING VERIFICATION CHALLENGES

A verification strategy should define an infrastructure that meets the following requirements in order to facilitate verification in the pre silicon environment:

1. Allows reuse between the Module, Chip and System level verification phases
2. Includes a robust verification metric which is reusable from one Phase to the next (Example: Module to Chip and System)
3. Minimal redundancy in both the checking and coverage in order to optimize the use of resources
4. Techniques for building a robust reusable verification test bench to accommodate different platforms (Verification, Validation-Includes ATE (Automatic Test Equipment), EVB (Evaluation Board) and System Verification.
5. Allows for both direct and random stimulus and defines their use
6. Challenges of building a robust reusable verification testbench?

> **Vertical Reuse between Module and Chip Verification**

The verification strategy for module (IP), chip and system level should complement one another and avoid redundancy. A redundancy results in simulation overhead with verifying same functionality and missing out on the corner case bugs. An IP verification strategy should verify the specifications. A chip level verifications strategy should verify all IP's are integrated correctly, meets the performance parameters of the SoC and finally ensure a flow for validation to verify silicon meets the design requirements. The system verification strategy ensures the use cases from the customer and architecture is verified prior the tapeout. This can result in verification components for IP, chip and system level test benches to be architected for integration and vertical reuse of verification components and verification intellectual property (VIP).

In order to meet design cycles set by market constraints, the specification and implementation phases of peripheral IP often occurs concurrently with the integration of the IP in a SoC. This poses a dilemma for verification engineers: should the effort be spent to create a standalone verification environment for the IP or should the SoC verification environment be used? Given the limited resource available for verification, creating the standalone environment can lead to delays in the SoC verification. In the parallel design and integration environment this predicament is especially pronounced. The need for standalone verification can not be overlooked, but at the same time, the SoC verification environment should not be ignored.

The need for standalone verification is simple: the cost-benefit equation for verification is favors a small design to a large design: the cost of verifying a two bit adder is significantly smaller than the cost of verifying a sixteen bit adder. The number of possible states that need to be verified in the two bit adder is factors smaller than the number of states that need to be verified in the sixteen bit adder. The same is true in the choice of verifying one IP at a time versus verifying many IP at the same time in the SoC environment. When verifying an IP standalone, the state space is reduced. This makes it possible, given time and machine resource constraints to explore more of the possible design state space and therefore increase coverage.

One problem with concentrating on standalone verification in the concurrent design and integration environment is the impact on SoC verification. While the majority of bugs found during the typical design cycle are IP related, SoC related defects are also a concern. The standalone environment may make certain assumptions about input signals that are incorrect or the integration of an IP may be incorrect. For example: an input may have been treated as active high during standalone verification, but may in fact be active low. The issue is how to maximize the effort spent on standalone verification (with its greater cost benefit ratio), while minimizing the impact on SoC verification.

One answer to this predicament is to take full advantage of Vertical Reuse. In general terms, vertical reuse stresses the development of verification components (VCs) that can be used at both the SoC and standalone environments. In order to ease the integration of these components into the SoC test bench, some constraints must be placed on there architecture of the test bench. A top down design style will facilitate development and should be used to reduce duplication. Duplication can drain resources during both the design stage of the VCs and during the simulation stage of the SoC test bench. For example, to separate IPs may require a common monitor for a bus. If both IPs develop separate monitors (that may be tweaked for that particular design), the development effort has been doubled. In addition the SoC test bench will require both of these monitors to be running, doubling the simulation effort for that monitor. In addition to inefficient use of resources, other issues can hamper virtual reuse if insufficient guidelines are given for development of verification components. Using a base, such as RVM™ from Synopsys, for all the VCs will allow them to effectively communicate even if they original came from different test benches. In addition, care must be taken in naming global variables, global defines and components. If a component is specific to one IP, it should be named as such, and not given a generic name. Wherever possible, the generic verification components such as reset, clocking and common busses should be identified and architected as early as possible. In name global variables and defines, it is very easy for collision to occur, and either have a failure in compilation or worse, a value that is not expected.

An example of an IP test bench is shown below in Figure I

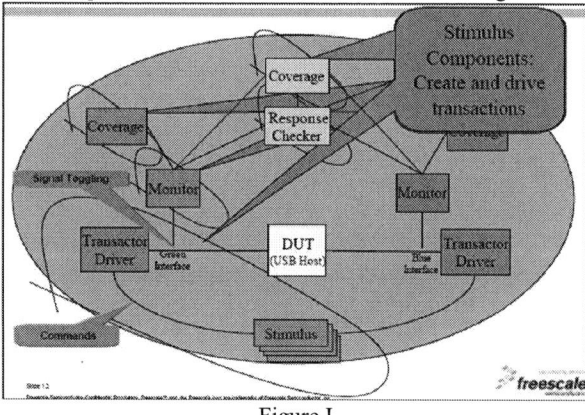

Figure I

There are other considerations that should be taken into account when developing verification components. In general, we can segment the verification components into two areas: a checking area which includes response checkers, assertion, coverage and monitors, and the stimulus area which includes stimulus and drivers.

➢ .Implementation of System Stimulus and software drivers

One area of particular interest is the design of system use cases. It can take many man years to develop test cases which are used to verify that the SoC can perform the required tasks. The verification team often times develops what appear to be paired down versions of the actual applications. The software

for SoC level verification is important for all aspects of the design process. Great care should be taken to develop the drivers in a top down fashion to maximize the reuse of the drivers in pre-silicon verification as well as post silicon validation. The difficulty faced with pre-silicon software development is that code development is very time consuming in an event based simulator. Tests that would take tens of milliseconds to run in real-time can take days to run and verify in simulation. One way to speed up simulation of the software is to use a hardware emulator or FPGA for development.

Emulation tools allow the SoC test bench team to rapidly develop system level drivers for the SoC level tests to use after the drivers have been integrated into the environment. Writing SoC level verification patterns using these drivers will verify the correct operation of the drivers, verify the correct operation of the design itself, and speed up the time taken to complete the verification tests. The drivers can then be ported with a high level of confidence to the post silicon environment very quickly.

Using the drivers in the full verification environment will also verify the correct operation of the emulation model. The emulation model can then be passed to the software teams to develop full application code for the SoC before silicon has been produced.

The integration of software drivers in the verification environment is very beneficial to all aspects of the design process. The simulation environment serves as a tool to verify the operation of the software that was developed rapidly in the emulation environment. This flow greatly speeds up the overall release process of the SoC as a product and creates a high level of confidence in all aspects of the design.

➢ **Implementation of Assertions, and response checkers**

As more peripherals are integrated in a SoC and various types of on-chip buses are introduced, it becomes a challenge that verifying these peripherals and buses interfaces conform to standard/custom protocols that are inflexible and timing relevant. In addition, with the increase of SoC scale and interactions between multiple modules, the cycles of simulation traces to identify the exact cause of an error goes up significantly in traditional debugging process.

Assertions are formal properties that describe design functionality or temporal relationships in a concise, unambiguous, and machine-executable way. Such formal semantics make it easier to express low-level signals behaviors such as interfaces protocols. In addition, assertions can be specified inline with RTL code, adding additional observability in the design and helping to detect and diagnose problems quickly. Assertions can also provide coverage metrics that can be combined with other forms of coverage, to confirm that the design has been thoroughly verified. Assertions not only provide a mechanism to verify design correctness but also a technique to ease debugging and measure verification quality.

Generally, design engineers write assertions to capture critical assumptions and design intentions during the implementation phase of the design. These assertions are implemented within the design and can be used to verify FSM state transitions, FIFO and decoding logic etc. Verification engineers focus on functional checks on external interfaces based on the design specifications. Protocols violation or specifications errors can be found in this way. Herein assertions are developed in separate files.

To solve the verification challenges for interfaces checking and making debug more efficient, assertions technique is adopted in SoC verification flow. There are assertions for protocols and chip level connectivity as well module level interfaces etc. The assertions adopting flow is shown in Figure II. After identifying the checking items to be covered by assertions, it might require considerable coding to describe complex design interfaces. Reusing libraries of pre-defined generic checkers (e.g. OVL) or assertion-based verification IP (AIP) will ease the coding effort as well speed up the adoption of assertions. Assertions that are package in reusable checkers are used in verifying the correctness of protocols and gathering functional coverage for protocols interfaces and connectivity

Debugging assertions requires a quite amount of effort as part of the adopting flow. While a failure flags, that could be: 1) a real design bug; 2) assertions coding error; 3) test bench bug. The correctness of assertions can be verified either through dynamic simulation or in formal verification tool. Assertions can be verified in formal verification tool first because formal technique is able to create stimulus using mathematical methods based on input assumptions, without the need to create testbench. It will examine all possible behaviors of the design and try to prove each assertion always holds true. For failure case, formal verification tool will produce a counter-example showing how the assertion is violated. However, there may be properties that can't be recognized or proved by formal tool, because formal analysis method requires high exponential memory.

Computing time complexity, also non-synthesizable assertions may not be supported by formal tool. These assertions are verified in dynamic simulation environment with a set of direct test cases or constraint-random stimulus.

Assertions provide coverage metrics to control the debugging process and determine that all assertions have been adequately verified. They also provide information that how well the design has been functionally tested. There are two types of coverage in assertions, one is structural coverage that measures each assertion and every path in an assertion is exercised. Another is user-specified functional coverage, which could be sequences coverage that is defined using cover property of SystemVerilog Assertions (SVA), or variables/expressions coverage modeled by functional coverage construct (e.g. SystemVerilog covergroup). They are beneficial to identify the functional tests holes and measure whether interesting scenarios such as specific data path or corner cases have been covered.

In addition to augment the coverage metrics for the SoC verification. assertions improve the quality of verification in other aspects such as documenting, debugging, and reusing etc Help clarify specifications and documentations.

Assertions provide an unambiguous way to describe design specifications in executable form, making it easier to capture mismatches between design and documentations, and reduce the misinterpretation of specifications.

Accelerate debugging of the design. An assertion in pair with failure message enables it to identify the problem quickly. Besides, assertions inside a design or on external interfaces make it fast to locate the exact cause of an error.

Reusable from module level to chip and system level.

Assertions verifying interface protocols can be easily customized as reusable checkers and integrated in chip and system level verification, to verify the interconnections and communications among blocks. Assertions within design can also be enabled in higher level integration.

However, since assertions can be used as checkers, avoiding duplicate checks in assertions and Response Checker (RSC) becomes a challenge for the IP verification flow, otherwise that will lead to redundancy and simulation overhead. Taking the example of Interrupt Request Controller (IRQC), which is a module that monitors 32 interrupts inputs and generates two composite interrupts as outputs, Figure III shows a brief block diagram for IRQC. IRQC supports edge/level sensitive mode and synchronization selectable for each interrupt source, also each interrupt can be high-true/low-true and individually enabled to either output. Besides, the state registers in IRQC provide visibility of all raw interrupt sources and unmasked sources pending for each interrupt output. The main checking items for IRQC include:

1. General registers read/write bus protocol on external interface
2. IRQC-specific bus error/wait behaviors, e.g.: error occurs when writing to a read-only register; wait cycles are asserted when reading/writing to a specific address.
3. Internal configuration/state registers read/write. These registers define the fundamental operation modes of the design and the states of them determine the design behaviors.
4. Correctness of two interrupts outputs based on inputs and the design configurations/state.

Both assertions (e.g. in SVA) and RSC (e.g. in Vera, Synopsys [TM]) are able to perform above checks, it's quite possible that the implementation in assertions and RSC overlap if different persons are responsible for code development without a global guide or monitor. Building a plan that separates the checks with assertions and RSC up front is highly required, which will act as a guideline for the development phases. The plan is created according to the advantages and limitations of different approaches. For example, assertions are best for verifying low-level signals relationships, whereas RSC are more appropriately to check transaction level behaviors that involve extensive data structures or complex algorithms. For the example of IRQC, the item1 and item2 are suitable to be developed with assertions because they target at interface protocol and temporal relationships; while it's better to use RSC to implement item3 and item4 since they relate to bus transactions and need to record design status in large time scales.

In addition, assertions can be used in coverage analysis as well as Coverage Objects. It will be helpful that EDA tools can automatically integrate all the coverage points with individual metric, which will help isolate functional holes.

With a clear plan and by using assertions in the appropriate verification scopes, assertion is an efficient technique to verify peripherals and on-chip buses protocols for the SoC, at the same time the gathering coverage is useful to determine how much the SoC is exercised and increase the confidence that the SoC can work properly.

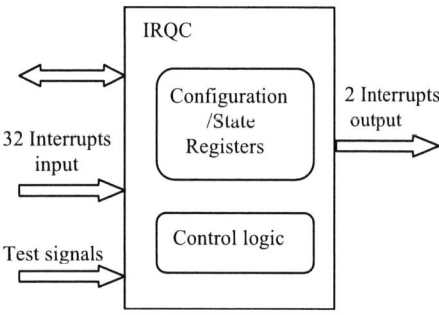

Figure II

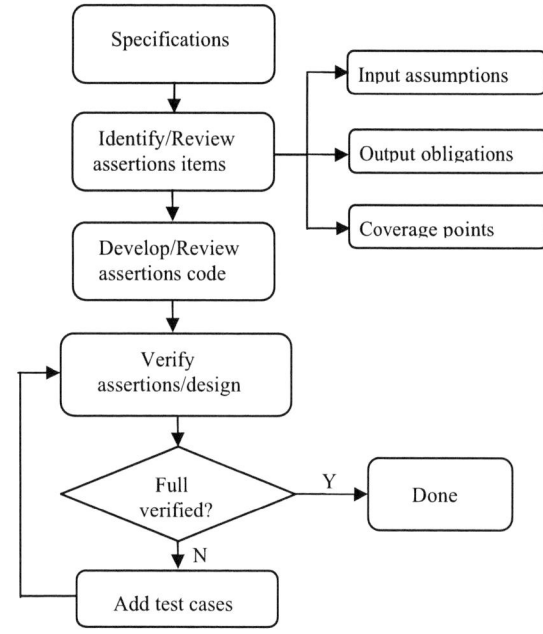

Figure III

> **Rerunning Functional Coverage for IP and SoC level**

IP's should be verified standalone independent of SoC integration. This means all feature sets of the IP should be verified standalone and the coverage metric (code coverage, functional coverage and assertion coverage) should be used to analyze the functional holes. The selection of module functionality and integration coverage should be identified from the specifications.

When porting coverage and response checkers to SoC any dip in functional coverage should be analyzed carefully.

The example below shows the data of functional coverage from a SoC with random verification. Modules with low functional coverage was analyzed and found to be due to integration at SoC. Example IP12 did not support all the feature sets at SoC level and constrained inputs of IP at integration resulted in not all functional coverage being achieved at SoC level.

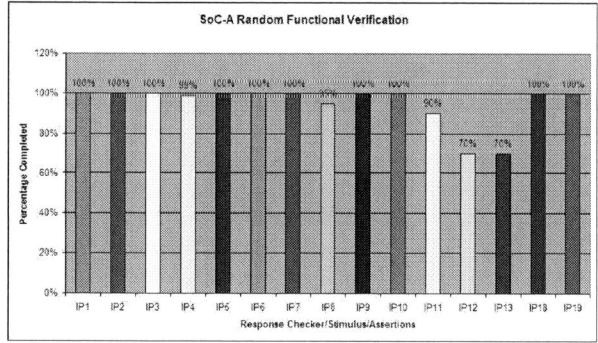

The recommendation is to analyze functional coverage holes at SoC level and understand why it could not be verified. Any uncovered functional hole may result in a functional bug for the IP.

IV. CONCLUSION

Some of the verification challenges for SoC described are related to planning, implementation. Vertical reuse helps in over head when moving from IP to SoC level verification review Stage. Verification metric in the forms of assertion, code coverage and functional coverage are all required to ensure there are no functional holes and verifying the corner cases prior to tapeout. The current EDA tools cannot integrate all the coverage bins into one composite bin, internal flows are used by industries to address this in verification planning.

Integrating software drivers in SoC verification will bridge the gap between hardware and software teams.

V. Authors

Noah Bamford (Ra8692@freescale.com) works for Freescale Semiconductor Ltd in Austin Texas, USA His interests are in the areas of testbench architecture, random verification, and low power verification.

Rekha K Bangalore(Rekha.Bangalore@freescale.com) works for Freescale Semiconductor Ltd In Austin Texas, USA. Her interests are in the areas of verification methodology, coverage, Design for Test and memory verification

Eric Chapman (Eric.Chapman@freescale.com) works for Freescale Semiconductor Ltd in Austin Texas, USA His interests are in the areas of testbench architecture and system verification methodology.

Hector Chavez (hector.chavez@freescale.com) works for Freescale Semiconductor Ltd in Guadalajara Mexico His interests are in the areas of IP and verification methodology.

Rajeev Dasari(Rajeev.Dasari@freescale.com) works for Freescale Semiconductor Ltd In Austin Texas, USA His interests are in the areas of testbench architecture, verification methodology and system level verification

Yinfang Lin(Linda.Lin@freescale.com) works for Freescale Semiconductor Ltd In Suzhou, China. Her interests are in the areas of IP SoC verification methodology.

Edgar Jimenez (Edgar.Jimenez@freescale.com) works. for Freescale Semiconductor Ltd in Guadalajara, Mexico. His interests are in the areas of IP and verification methodology.

VI. References

System-On-A-Chip Verification: Methodology and Techniques By Prakash Rashinkar, Peter Paterson, Leena Singh

Modeling IP Responses in Testcase Generation for Systems-on-Chip Verification (MTV2003)
 By Mrinal Bose, Mark H. Nodine, William R. Jurasz, Jr. Vlad Zavadsky, Arvind Chodavadia, Lincoln R. Nunes
Motorola,Inc.,Austin,TX,

A hardware and software monitor for high-level system-on-chip verification by El Shobaki, M.; Lindh, L.; (Quality Electronic Design, International Symposium March 26-28th 2001).

System-on-chip:reuse and integration Saleh, R.; Wilton, S.; Mirabbasi, S.; Hu, A.; Greenstreet, M.; Lemieux, G.; Pande, P.P.; Grecu, C.; Ivanov, A.;Proceedings of IEEE, June 2006, Volume 94

Comparison of verification methodologies for datapath testing, Iyer V.V. Microprocessor Test and Verification: Common Challenges and Solutions, 2003. Proceedings. 4th International Workshop on May 20-30 2003

VII. Acknowledgements

The authors would like to thank Pastukhov Sergey (Sergey.Pastukhov@telma.ru) from Telma Soft Ltd. Sergey participated in the verification methodology review and implementation of functional coverage.

SECTION 3: ARCHITECTURAL AND DESIGN ISSUES

Workload Slicing For Characterizing New Features in High Performance Microprocessors

Hassan Al-Sukhni, David Lindberg, James Holt, Michele Reese
{Hassan.Alsukhni, David.Lindberg, Jim.Holt, Michele.Reese}@freescale.com
Freescale Semiconductor, Inc.
7700 W. Parmer Ln, Austin, TX 78729

Abstract

Detailed pre-silicon analysis and validation of new features incorporated in high performance microprocessors often requires the use of RTL and gate-level models because accurate higher-level models of such features are not available. While the results of such studies can be very valuable, this approach is both slow and complex, resulting in extreme constraints on the maximum size of workloads that can be studied. Although micro-benchmarks can be used for this purpose, it is also desirable to characterize important new features using real-world workloads. Thus, a flow for characterizing new features using detailed RTL models requires workload sampling techniques. Existing workload sampling techniques use predefined metrics to identify representative samples of the workload, but these metrics may not encompass the requirements of a given study. A more flexible approach towards specifying arbitrary metrics is needed to enable extraction of representative samples of workloads that exercise specific features of the microprocessor. In this paper, we present Workload Slicing Flow as a set of tools that enables the selection of representative workload slices satisfying a set of metrics and constraints. The use of the flow is illustrated by selecting slices for power-characterization of the floating-point unit of a research microprocessor. The selected slices represent 4% of the original workload size, and result in power estimates within 2% of the full workload power estimates.

1. Introduction

Power consumption is becoming a first-order design concern of microprocessors. As such, power estimation is needed during the early design stages to evaluate alternative design options, and during later design stages to validate meeting the power targets and constraints for a specific application domain. This concern with microprocessor power is a direct result of the increasing demand for lowering the power consumption of systems due to several factors that include: (1) reliability (2) cost of packaging and cooling, and (3) longer operation-time-between-charges of battery-operated personal computing devices and wireless communication systems [1].

One challenge is that models with different levels of abstraction are used to characterize micro-processors during the design cycle. Low-level models of the microprocessor are often used to characterize power consumption of new features because of the lack of higher-level models for the new features, or because higher-level models do not provide sufficient accuracy. Figure 1 illustrates typical abstraction levels at which power consumption can be characterized.

The power characterization tools for low-level models of complex designs such as high-performance microprocessors are very costly to run in terms of both time and space. Furthermore, lower abstraction levels require more complex models, which results in longer simulation times, and larger output sizes. As such, the stimuli that can be used at these low abstraction levels is subject to size constraints.

Several techniques can be used to generate small-sized stimulus for power characterizations with low-level models of microprocessors, including micro-benchmarks, and benchmark sampling. Unfortunately, these techniques suffer a few limitations, which make it desirable to develop additional capabilities in the stimulus generation tools.

Micro-benchmarks are good for targeted characterization and testing of specific functionality of new features. However, micro-benchmarks do not scale well to higher-level models because they usually are not representative of workloads of interest, and hence cannot be used for estimating their power consumption.

Sampling techniques have been successfully used in architectural and micro-architectural

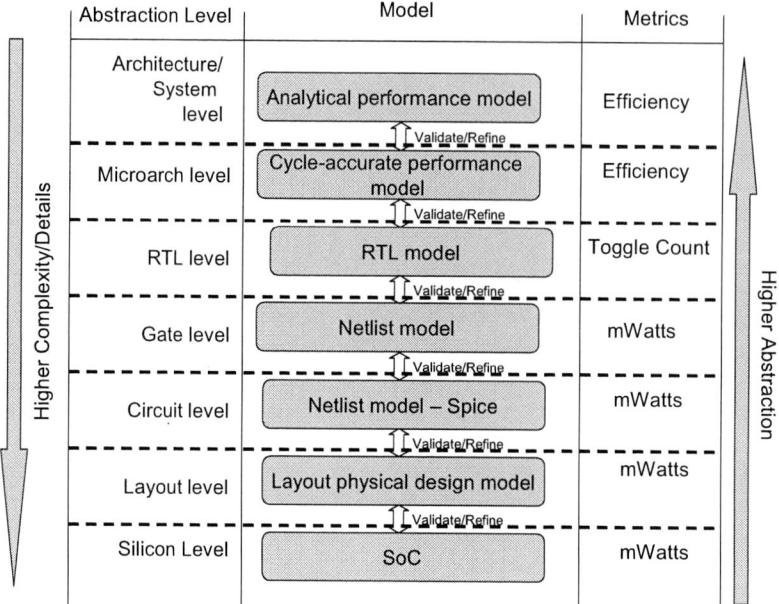

Figure 1 Power estimation abstraction levels.

design exploration studies [2]. Tools like SimPoint [3] use basic-block execution counts to identify phases of program execution, and after that, use a clustering algorithm to find a set of representative samples that can be executed in lieu of the program. In contrast, ad-hoc sampling techniques such as fast-forwarding, reduced input sets, etc. have been demonstrated to be less effective in producing samples that are representative of their programs [4].

Existing sampling techniques when used to generate workloads for power characterization suffer two main problems. The first is that the available tools for sampling do not offer flexibility in specifying metrics of interest that the samples should exhibit. For example, characterizing the power consumption of a floating point unit in a micro-processor requires workloads that include floating point instructions. Samples of a program from regions of code that do not include such instructions are not of interest for this characterization. As such, a sampling tool that provides flexibility in specifying the sampling metrics and constraints is necessary for effective characterization of specific features of a hardware design.

The second problem with existing statistical sampling techniques is that they assume that the sample size is large enough to allow the initial state of the micro-architecture to be ignored; therefore these techniques do not account for

warm-up of the samples. Unfortunately, our experiments indicate that the micro-architecture state can influence the characterization to a large degree, considering the workload length constraint on stimulus used for power characterization with low-level models. For example, executing a conditional branch instruction in an out-of-order microprocessor after its branch predictor has warmed-up can result in a completely different sequence of instructions than if the branch predictor has not been warmed-up. This difference can be on the order of tens of instructions, and if the whole sample consists of tens of instructions (which is a typical size at a gate-level simulation), then this may result in a large margin of error in the characterization.

This paper presents *workload slicing* as a sampling technique that provides innovative capabilities for specifying the sampling metrics, and overcomes the above problems associated with existing sampling tools for simulation at the low-level abstraction models. Workload slicing examines a large code sample (or program) to find short representative regions of the sample, called *slices*, which collectively represent the sample for a given *metric* (metrics will be further explained in Section 2). A slice includes initial state information, such as the state of the general purpose registers, and sufficient memory initialization of code and data sections. The next section explains the slicing approach and slicing

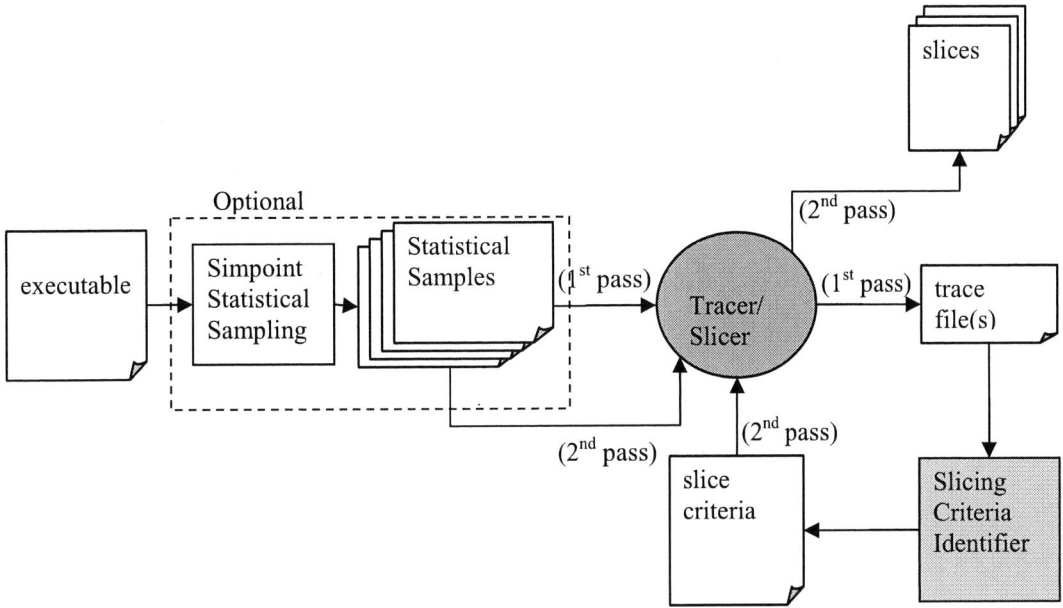

Figure 2 Workload slicing flow.

flow. Section 3 presents validation experiments for workload slicing. Finally, Section 4 summarizes this paper, and gives future work suggestions and ideas.

2. Approach

Workload slicing is carried out by processing the workload in two passes. Figure 2 illustrates the workload slicing flow, and the tools used in the flow. The first pass identifies slicing criteria, while the second pass produces the slices. In the first pass, the workload executable is traced using an Instruction Set Simulator (ISS), and the trace is annotated with information that is extracted from a model of micro-architecture aspects of interest, like a cache model. This is done using the Tracer/Slicer in Figure 2. In addition to tracing the workload, the Tracer annotates the resulting traces with information that is used by other tools of the flow. Examples of such annotations are cache hit/miss information. The traces in this pass are then fed to a tool referred to in Figure 3 as Slicing Criteria Identifier (SCI), which identifies slicing points based on given metrics of interest, and constraints. In the second pass, the workload is run again to generate slices based on the identified criteria. This is done by the Slicer, which also generates micro-architecture warm-up prelude, as discussed in Section 2.3.

Workload slicing carries out three main tasks: statistical sampling, slicing, and micro-

architecture warm-up prelude generation. These tasks are discussed in the following sections.

2.1. Statistical Sampling

The first task in the slicing process is statistical sampling, illustrated in Figure 3(a). Statistical sampling is an optional task and is used to reduce the workload size for the subsequent tasks. A statistical sampling tool, like Simpoint, is used to find samples that represent the different phases of the workload based on the execution counts of its basic blocks. Each sample is associated with a weight that indicates the percentage of its basic blocks execution to the overall workload basic blocks execution. In the subsequent tasks, each sample's metric of interest is estimated, for example its power consumption. The weights associated with each sample can then be used to estimate the workload's overall metric of interest, e.g. its total power consumption.

2.2. Slicing

The second task, illustrated in Figure 3(b), involves using the SCI to identify representative slices within the traced samples. The slices represent the samples based on a specific metric (or set of metrics), and satisfy a set of constraints. This paper illustrates the use of the slicing flow for finding slices to enable studying the floating point unit in a research microprocessor. For this example, the slicing criteria include the density of floating point

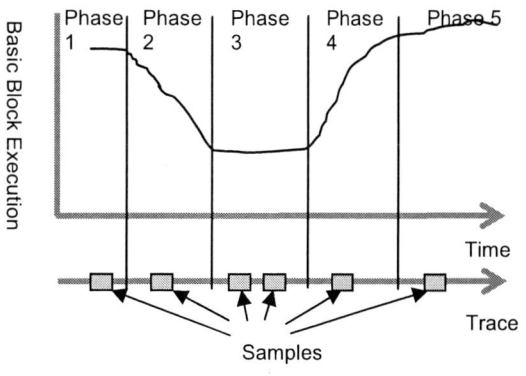
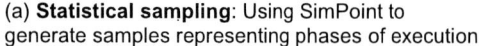

(a) **Statistical sampling**: Using SimPoint to generate samples representing phases of execution

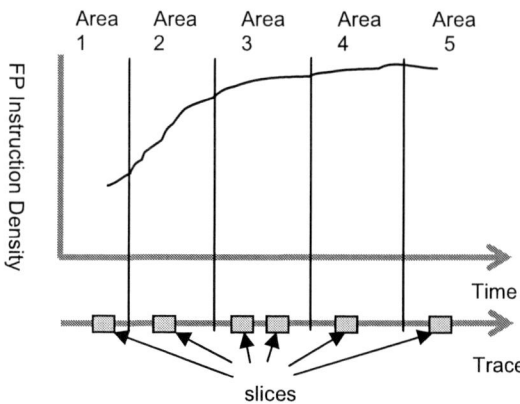

(b) **Slicing**: Using a slicing tool to generate slices representing phase samples

Figure 3 Statistical sampling and slicing illustration.

instructions in the code as a metric, as well as a constraint to slice at loop boundaries.

2.3. Micro-architecture warm-up

The final task is to generate the identified slices from the samples, and add the necessary warm-up of the micro-architecture to the slices. This task is carried out by the Tracer shown in Figure 2, and it consists of three sub-tasks: initial state warm-loading, cache warming, and branch-predictor warming. These sub-tasks are explained next.

2.3.1. Initial state warm-loading

Many test benches provide the ability to specify initial state values for the different architected resources, like registers. The values of such resources at the beginning of the slice are collected at the slicing point by the Slicer, and are output in a format understood by the test bench as part of the final slice. If the test bench does not provide resource warm-loading facilities, then code execution can be used to initialize the required resources to the desired initial state.

2.3.2. Cache warming prelude

Cache hits and misses can affect the execution of slices to a large degree, especially in multi-threaded micro-architectures, where cache misses are typically used for thread switching. Cache misses influence the execution time of workloads. As such, cache warming is necessary for slices to execute in a similar fashion to that of executing the sliced code within the context of full workload execution. In order to warm the caches, a prelude code section is added to the slice. The prelude code includes prefetch

instructions (supported by most architectures) for addresses that would hit in the context of full workload execution. These addresses are collected from the cache activity annotations that the Tracer added to the trace in the first pass.

2.3.3. Branch-predictor warming

Branch predictor warming is necessary for micro-architectures that support speculative execution. This is due to constraints on the length of slices that can run on low-level models. The branch predictor is likely to make different predictions for branches contained within a slice when the slice is run in isolation as compared to when the same sequence of instructions is run as part of the full workload. This difference in predictions can result in executing different sequences of instructions during running the slice.

Identifying branch predictions during the tracing pass requires a detailed model of the micro-architecture. Such a detailed model can slow the tracing task significantly. In order to overcome this problem, an approximate model of the branch predictor can be used during the tracing pass. The predictions made by the branch predictor model are annotated into the resulting trace. These annotations are used by the Slicer to add hints to the branch instructions of the slice. Several architectures provide capabilities for communicating such hints to the micro-architecture, causing the branch predictor to make a prediction similar to the one that was done during the full workload run.

1550-4093/07 $25.00 © 2007 IEEE

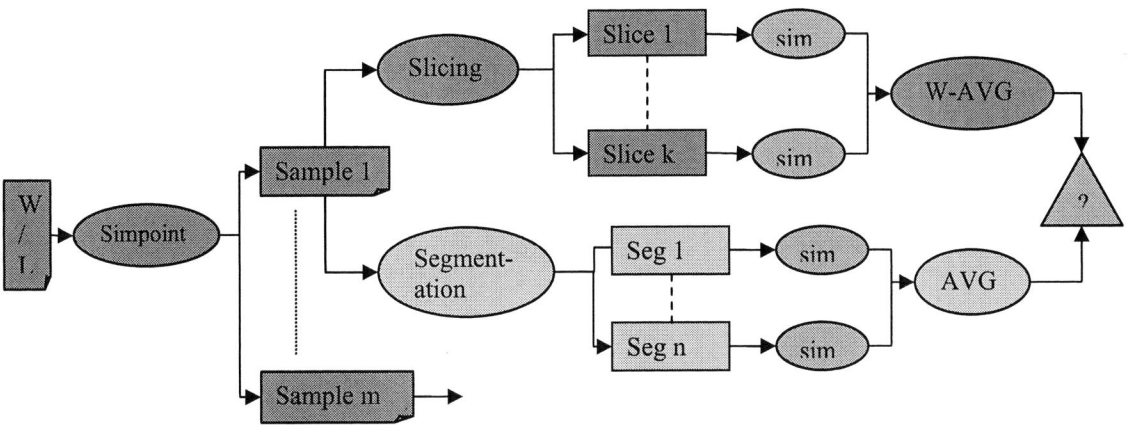

Figure 4 Workload slicing validation methodology.

3. Validation Experiment

3.1. Methodology

In order to evaluate workload slicing, an experiment was designed to compare the power consumption of a floating-point unit of a research microprocessor estimated using slices to the power consumption of the sliced workload. The benchmark Whetstone [5] was used as a workload input to the slicing flow. Simpoint was used to sample the benchmark, with a fixed sample size of one million instructions. One of the resulting samples was sliced using the slicing flow to identify its representative slices, as illustrated in Figure 4. The representative slices were simulated on a gate-level model, to generate bit switching information, which was then fed to a tool that estimated the power consumption of each slice. The slices' power consumptions were used to estimate the sample's power consumption.

Finding the power consumption of the floating point unit for the overall sample was not practical due to its large size. Therefore, the sample was segmented into equal-sized segments. All the segments were then individually run through the power measurement flow described above. The sample's power was then computed as the average of its segments' power consumption.

The sample consisted of mainly two loops. Loop 1 accounted for about 86% of the sample, while loop 2 accounted for about 14% of the sample size. The criteria given to the SCI included slicing at loop boundary, density of floating point instructions, and a slice length of about 100 instructions. In addition, the SCI was instructed to generate approximately 100 slices for each of the two loops in order to account for data-dependent power readings (in this case, a slice represents a loop iteration). The SCI generated 100 slices for loop 1 and 104 slices for loop 2, along with weights for each slice. The weights represent the instructions count percentage of each slice. The resulting slices' power consumptions were estimated as explained earlier. The total sample power consumption was then computed as a weighted average of the power consumed by each slice.

3.2. Results

Figure 5 depicts the power consumption for each slice representing loop 1 of Whetstone. The difference in power consumed by the different slices is mainly due to difference in processed data of each slice. The power ranges from 0.481 Watts to about 0.518 Watts, with an average weighted power of 0.499 Watts. The similarity in consumed power between slices is not surprising since the same code is executed in all the slices.

Figure 6 depicts the power consumption for each slice representing loop 2 of the studied Whetstone sample. Again, the difference in power consumed by the different slices is mainly due to difference in processed data of each slice. Here, the power ranges from 0.485 Watts to 0.595 Watts, with an average weighted power of 0.513 Watts. The power consumed by the different slices is not as similar as that seen for loop 1. Analysis revealed that this can be attributed to different control paths in the code for loop 2, which resulted in different instruction sequences between the different slices. The overall sample's weighted power is 0.501 Watts.

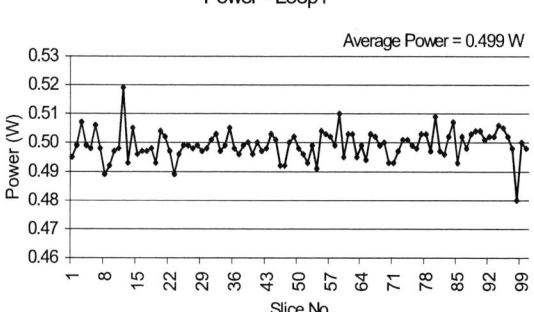

Figure 5 Power consumption of loop 1.

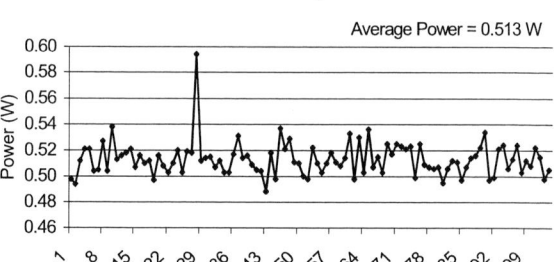

Figure 6 Power consumption of loop 2.

Finally, Figure 7 shows the power consumed by each segment of the studied sample of Whetstone. The power consumed by the individual segments ranges from 0.441 Watts to 0.641 Watts. The average power consumed by all the segments is 0.492 Watts. On the other hand the estimated power consumed by this sample computed from the power of loop 1 and loop 2 above is 0.501 Watts.

The main result of this experiment is that running slices that are about 4% of the total sample size resulted in power estimation within 2% of the power consumption found by running all the segments of the sample.

4. Summary

Power estimation and characterization of new features of high-performance micro-processors is becoming a first-order design concern. The ability to perform power estimation and characterization is limited by available models of these features. These models present constraints

on the size of stimulus that can be used.

Micro-benchmarks and benchmark statistical sampling are not sufficient to produce representative workloads for power estimation, because of scalability, or the lack of flexibility in the metrics and constraints that the sampling tools provide. As such, workload slicing was presented here to produce representative small slices that can be run in lieu of long workloads.

In addition to the requirement for flexibility in the criteria used for slicing (both metrics and constraints), it was found that micro-architecture warm-up is necessary for short slices to be used in estimating the power consumed by the long benchmarks. Two main warm-up problems were discussed: cache warming, and speculative execution due to branch predictor warm-up. Mechanisms for warming-up the micro-architecture state were presented.

The evaluation experiment presented in this paper illustrated that about 4% of a particular

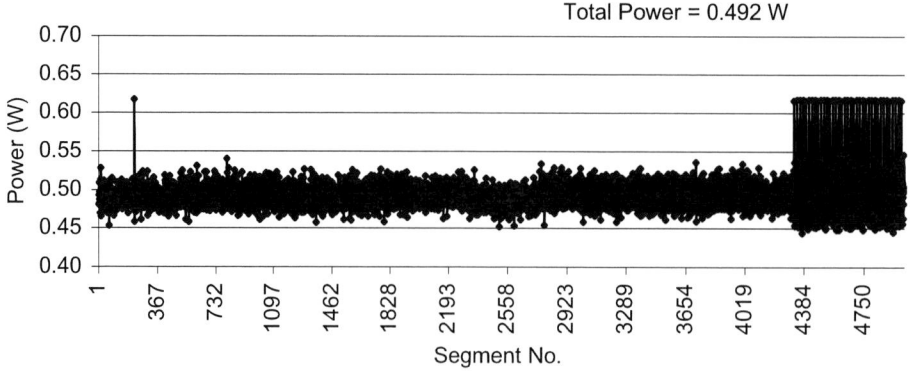

Figure 7 Segments power consumption of the sliced sample.

1550-4093/07 $25.00 © 2007 IEEE

workload was sufficient for estimating the power consumption to with 2% accuracy. This result is an encouraging one, motivating us to pursue evaluating workload slicing.

Experiments are needed to further evaluate this proposed approach. The presented experiment does not provide statistical significance, and as such, more workloads of different domains need to be considered to further establish confidence in the approach. Furthermore, additional validation is needed beyond the segment level, and up to the full benchmark level.

5. References

[1]   E. Macii, M. Pedram, and F. Somenzi, "High-level Power Modeling, Estimation, and Optimization," *IEEE Transaction on Computer-Aided Design of Integrated Circuits and Systems*, vol. 17, pp. 1061-1079, 1998.

[2]   H. F. Al-Sukhni, J. C. Holt, and D. A. Connors, "Improved Stride Prefetching using Extrinsic Stream Characteristics," presented at IEEE International Symposium on Performance Analysis of Systems and Software, Austin, Texas, 2006.

[3]   E. Perelman, G. Hamerly, M. V. Biesbrouck, T. Sherwood, and B. Calder, "Using SimPoint for accurate and efficient simulation " in *Proceedings of the 2003 ACM SIGMETRICS international conference on Measurement and modeling of computer systems* San Diego, CA, USA ACM Press, 2003 pp. 318-319

[4]   J. J. Yi, S. V. Kodakara, R. Sendag, D. J. Lilja, and D. M. Hawkins, "Characterizing and Comparing Prevailing Simulation Techniques," presented at International Symposium on High-Performance Computer Architecture, 2005.

[5]   Netlib.org, "The Whetstone Benchmark," http://www.netlib.org/benchmark.

Deep vs. Shallow, Kernel vs. Language – What is Better for Heterogeneous Modeling in SystemC?

Hiren D. Patel and Sandeep K. Shukla
Center for Embedded Systems for Critical Applications
Virginia Polytechnic Institute and State University
Blacksburg, Virginia, 24061, USA
{hiren, shukla}@vt.edu

ABSTRACT

It is common for large designs to have heterogeneous components interacting with each other. These components often follow a particular model of computation such as controllers modeled using state machines, signal processing filters modeled as data flow and event-based components using discrete-event. Hence, there are several academic and industrial attempts at incorporating heterogeneity into the design flow, primarily in system level design languages and frameworks for modeling and simulation. A variety of attempts are proposed such as extending simulation kernels for existing frameworks and simply using language constructs to mimic other models of computation. However, the benefit of one over the other is not apparent to the designer and thus not clear which of the two is a better strategy for EDA tools to integrate. In this paper we argue whether Deep heterogeneity (kernel-level) or Shallow heterogeneity (language-level) is a suitable strategy for introducing heterogeneity in system level design languages and frameworks.

1. INTRODUCTION

In order to manage the heterogeneity and complexity demands for efficient system design, many different and overlapping methodologies have developed themselves into system level design languages and frameworks (SLDLs) for modeling and simulation. Some examples of interesting methodologies recently proposed are by SystemC [20] for raising the level of abstraction from register-transfer language (RTL) based modeling and simulation to transaction-level and beyond, SystemC-H [5] for introducing heterogeneity in SystemC, Ptolemy II [18] for heterogeneous behavioral hierarchy, Metropolis [9] for their refinement-based and meta-modeling approach, EWD for visual meta-modeling [2], SML-sys [19] for their functional paradigm based system design and so on.

Even though all these SLDLs and frameworks facilitate heterogeneity in modeling complex systems, they are realized in different manners. SystemC being developed since its original almost RTL level implementation to today's transaction-level modeling language has side stepped the issue largely except for some academic attempts at making heterogeneous modeling extensions to it [14, 16]. However, the inherent model of computation (MoC) embedded in the original SystemC kernel remains the same in the language standard [20]. Interestingly, the discrete-event (DE) simulation is a powerful MoC and hence one can retrofit many other MoCs on top of it. SystemC being a class library of C++ has a higher flexibility at the syntax level as well. Therefore, two possibilities arise while considering heterogeneous modeling with SystemC, and understanding the relative pros and cons of these is of importance to CAD designers.

The capability of expressing heterogeneity is an essential ingredient for raising the level of abstraction of an SLDL. For example,

controller behaviors may be expressed using the finite state machine (FSM) MoC and that of the DSP cores as a data flow (DF) MoC. An MoC describes the manner in which the computation occurs and the way in which the communication proceeds with components within that MoC. Examples of other MoCs are communicating sequential processes (CSP) [11], synchronous data flow (SDF) [12, 1], continuous time (CT), and discrete-event (DE) [20, 6, 14, 16, 18, 13]. A natural extension to heterogeneity is modeling with heterogeneous behavior hierarchy. Such modeling enables embedding of already existing component models inside a larger model where every component possibly follows a different MoC. This sort of design requires hierarchical composition of heterogeneous components.

1.1 Implementing Heterogeneity

We propose that SLDLs should possess the capability for heterogeneity and heterogeneous behavioral hierarchy (HBH). As mentioned before, there are two methods of providing capabilities for expressing HBH, one being at the kernel-level, which we term *Deep* heterogeneity and the other being at the language-level, which is *Shallow* heterogeneity. The capability of heterogeneity may be ingrained deep into the simulation kernel or superficially at the linguistic level, where some semantic mapping onto a single MoC kernel can be used for simulation. We interchangeably use kernel-level for deep and language-level for shallow throughout this paper. Examples of expressing HBH at the kernel-level are SystemC-H [5], SystemC-AMS [7] and Ptolemy II [18], which implement MoC-specific kernels to allow for heterogeneity in their language and framework. Authors of [10] also attempt to integrate MoCs into SystemC at the language-level. On the other hand, examples of expressing HBH at the language-level is already commonly used in SLDLs. In SystemC for example, a controller component of a design implements an FSM, which otherwise when using SystemC-H or Ptolemy II requires using the FSM MoC. Similarly, authors of [16, 8] indicate that with effective programming skills a user may introduce many other MoCs using just SystemC's capabilities of primitive and hierarchical channels based on the DE MoC, thus more examples of language-level expressiveness of HBH. However, the difficulty level is high when considering specific MoCs not entirely suitable for the DE semantics. For example, the continuous-time MoC and the simulation of analog components requires a large library of differential equation solvers which with just SystemC is not possible. Hence the SystemC-AMS extensions [21, 7].

In doing the classification, we analyze the two approaches and argue which of the two methods is a better strategy for introducing heterogeneity in SLDLs. Our experimental SLDL is SystemC [20] due to its open-source nature and its strong industrial traction. We

1550-4093/07 $25.00 © 2007 IEEE

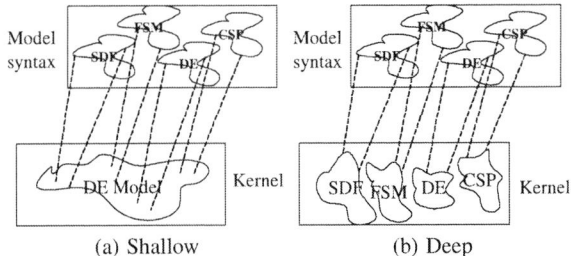

Figure 1: Deep vs. Shallow Heterogeneity

evaluate the best strategy by discussing restrictions imposed by SystemC's reference implementation in cleanly integrating extensions at the kernel-level, the overhead in C++ programming skills required in modeling designs other than those within the DE MoC, and the presentation and intuitiveness of the coding requirements. We provide some examples that show code snippets of implementations using SystemC's reference implementation along with SystemC-H's kernel-level extensions.

2. MAIN CONTRIBUTION

It is conceivable that multi-MoC support for heterogeneous system level modeling can be provided in two ways:

1. syntactic extensions at the modeling language-level only (shallow). In C++, syntactic extensions amount to MACROs and class libraries.
2. some syntactic extensions with the addition of the simulation/synthesis level support for multi-MoCs (deep).

Figure 1 shows that in the first case, every model maps to a fixed MoC while simulating (such as described by Grotker et al. [8]) and in the second case the simulation engine or synthesis engine are made aware of the MoCs through implementation of multi-MoC kernels. In this paper, we take a critical look at these possibilities and argue which of the two is a more desirable CAD strategy.

3. RELATED WORK

In this section, we briefly discuss some of the popular frameworks and languages that allow expressing design components using multiple MoCs. SystemC-H is especially important because the comparison of examples shown in Section 5 use SystemC's reference DE kernel and SystemC-H's extensions for the FSM, SDF and CSP MoCs.

3.1 Ptolemy II

One of the renowned promoters of heterogeneous and hierarchical system design is the Ptolemy Group at U. C. Berkeley [18]. Their experimentation with heterogeneity and hierarchy began with the Ptolemy Classic project where they introduce the FSM and SDF MoCs. However, this project was later abandoned and rejuvenated as the Ptolemy II project which is a Java based implementation of their multi-MoC framework. Ptolemy II follows an actor-oriented approach for modeling designs with a Java-based graphical user interface through which designers can "drag-and-drop" actors to construct their models. The two main types of actors are atomic actors and composite actors. The former describes an atomic unit of computation and the latter describes a medium through which behavioral hierarchy is possible. Ptolemy II's directors encapsulate the MoC behavior and simulate the model. A model or

component that follows a particular director is said to be a part of that MoC's domain. Few of the domains implemented in Ptolemy II are component interaction, communicating sequential processes, continuous time, discrete-event, finite state machine, process networks, data flow, and synchronous data flow. Ptolemy II clearly falls into the category which implements deep heterogeneity because its simulation is driven by specific multi-MoC kernels.

3.2 SystemC-H

An experimental prototype developed to study heterogeneity in SystemC labeled SystemC-H [5] introduces three MoCs interoperable with SystemC 2.0.1. The three MoCs are FSM, SDF and communicating sequential processes (CSP). SystemC-H imposes stylistic guidelines to increase the modeling fidelity [16] of SystemC. They define fidelity as the capability of the framework to model a theoretical MoC. Details with code examples and descriptions are given in [14, 16]. SystemC-H also shows that provided the simulation kernel supports heterogeneity, certain optimization for simulation efficiency are possible. They show examples of the SDF MoC increasing simulation efficiency [14, 16] by approximately 50%. Though SystemC-H supported heterogeneity, it lacked behavioral hierarchy not allowing designers to hierarchically compose designs with varying MoCs. In [17], the authors enable the important quality of heterogeneous behavioral hierarchy in SystemC by deep extensions to the kernel. They also introduce additional syntax to describe models to the extended kernels. Therefore, SystemC-H also implements deep heterogeneity.

3.3 YAPI

YAPI is a C++-based run-time signal processing programming interface to construct models using the Kahn process networks (KPN) MoC [3]. The purpose of YAPI is to primarily allow re-usability of signal processing applications for hardware and software codesigns. This programming interface follows a homogeneous MoC, primarily the KPN MoC. The main disadvantage for systems designers is that systems nowadays are heterogeneous with more behavioral components than just signal processing cores. We feel that the purpose of YAPI is specific to signal processing applications, hence, on it its own, it does not suffice the need for a multi-MoC framework for heterogeneous and hierarchical designs. YAPI therefore supports shallow heterogeneity. KPN-based models are simulated by the YAPI run-time library but every other MoC requires modeling using C++.

3.4 Metropolis

Another U.C. Berkeley group works on a project called Metropolis [9], whose purpose is again directed towards the design, verification and synthesis of embedded software. However, the approach they employ is different than that of Ptolemy II's. Metropolis has a notion of a meta-model as a set of abstract classes that can be derived to model various communication and computation semantics. The basic modeling elements in Metropolis are processes, ports, media, quantity manager and state media. Processes are atomic elements describing computations in its own thread of execution that communicate through ports. The ports are interfaced using media and the quantity manager enforces constraints on whether the process should be scheduled for execution or not. The quantity manager communicates through a special medium called state media. Defining the communication elements such as the media, determines the MoC the platform follows. Heterogeneity is achieved by differing the implementation of these media. Unfortunately, heterogeneous behavioral hierarchy is not possible in Metropolis because the heterogeneous components require a communication me-

1550-4093/07 $25.00 © 2007 IEEE

dia/medium between the two to transfer tokens causing them to be at the same level of hierarchy. We classify Metropolis implementing shallow heterogeneity because even though Metropolis allows definition of medias for different MoCs, they are still simulated using the same PN-based kernel.

3.5 HET-SC

The heterogeneous library for SystemC [10] is a library of SystemC channels with synchronization primitives allowing heterogeneous modeling with MoCs such as the process network, synchronous reactive, synchronous data flow and communicating sequential processes. This is an excellent example of shallow heterogeneity with SystemC. However, the source for this project is unavailable thus we cannot compare kernel-level solutions against this implementation for shallow heterogeneity. Instead we employ our own methodology in modeling shallow heterogeneity in SystemC.

3.6 SystemC-AMS

The SystemC-AMS [7, 21] project inspired by the VHDL-AMS project, introduces a continuous-time MoC with the capability of modeling and simulating analog components such as RF, wireless, and digital signal processing applications. Their extension brings forth the continuous-time, data flow, and synchronous data flow MoCs into SystemC. SystemC-AMS employs a layered approach. The first layer is the user layer that provides descriptive methods to create continuous-time models. The second is the interface layer that interacts with the solvers and finally the synchronization layer which takes care of the interaction between the continuous-time and DE MoCs. The approach employed by SystemC-AMS is of deep heterogeneity.

4. SHALLOW OR DEEP HBH

Augmenting SLDLs with HBH allows for high modeling fidelity [14] in system level modeling. Fidelity is defined as the capability of the framework to faithfully model an MoC. We do not delve into a formalization of what it means to *formally* represent an MoC, but rather appeal to intuitive sense of readers for the informal meaning of the statement. SystemC-H [5] shows one of the first contributions to increasing modeling fidelity by promoting the idea of deep heterogeneity in SystemC via the introduction of MoC-based kernel extensions. However, the classification of heterogeneity through language-level or kernel-level support was not clearly distinguished nor discussed in those works.

Heterogeneous behavioral hierarchy can also be supported at the kernel and language-levels. At the language-level, with the aid of a library of primitive and hierarchical channels in SystemC, a user can construct components and embed them with other components realized by different MoCs. During simulation, this HBH model is flattened and the DE simulation kernel treats all SystemC processes at the same level of hierarchy identifying which processes to simulate next based on the events that are notified. Simulation of the same HBH model with deep heterogeneity is done differently, whereby each level of hierarchy has its own instance of the particular MoC's kernel [15]. Hence, the model follows strict execution semantics defined for the composition of different MoCs.

4.1 The Deep: Extending SystemC's Kernel

Kernel-level support for heterogeneity suggests that there are MoC-specific kernels responsible for simulating models described by the MoCs. Note that kernel-level support is not totally independent of language-level support because a specific MoC-kernel may also require the addition of constructs to the language. This

has been the primary focus of the work in [14, 16], where extensions for SystemC with the SDF, FSM and CSP MoCs at the kernel and language-level are discussed. We say at both levels because language-level changes are necessary in order for the extended kernels to infer information specified by the user, hence also extending the set of language constructs of the SLDL. More recent work in [17, 15] explores heterogeneous behavioral hierarchy again with the SDF and FSM MoCs.

From SystemC's DE kernel [14, 16, 15], we feel that the authors of the reference implementation did not foresee users attempting to extend at the kernel-level. Augmenting SystemC [20] with capabilities of HBH exposed limitations with SystemC's kernel. These limitations in turn impose constraints on how deep heterogeneity is introduced across all MoCs and behavioral hierarchy for the DE MoC. To better understand these constraints we first briefly describe SystemC DE kernel's singleton pattern implementation. Note that SystemC's DE is one implementation of the DE semantics and there can be various different implementations. SystemC requires the user to associate the modeled behavior with particular SystemC processes. The two types of SystemC processes are SC_METHOD and SC_THREAD. The class responsible for simulating the model is implemented in sc_simcontext, which interacts with the coroutine classes. The sc_simcontext when starting the simulation creates a static instance of sc_simcontext which is used as the default context. The default context is assigned as a static object, which is invoked from the main entry function call sc_start(). The static nature of sc_simcontext restricts another instance to replace the default context, thus making SystemC's DE kernel follow a singleton design pattern [14]. This means that a single global or static instance of the simulation class holds the default context.

When extending SystemC, the existing SystemC reference implementation needs to remain unaltered, therefore the singleton pattern design of SystemC's DE kernel imposes hard and fast restrictions on how to introduce HBH with SystemC as well as how to integrate extensions to it. Due to the restriction imposed by SystemC's kernel, it is not possible to have two concurrently executing instances of SystemC's DE kernel. Even though the simulation functionality is tidily encapsulated in class sc_simcontext, two instantiations of sc_simcontext only results in loss of context information. To introduce HBH, we require the capability of having multiple instances of the DE kernel such that one instance may be embedded within another, each instance may further be embedded in other MoCs, and so on. As of now, we do not possess the capability to explore behavioral hierarchy with SystemC's DE kernel, but we have a plan set aside for our future work to investigate these necessary hierarchical compositions.

Another artifact of the SystemC kernel's implementation is that all SystemC processes are treated at the same level of hierarchy, in essence flattening whatever hierarchical structure presented in the model [15]. The SystemC processes as seen in class sc_simcontext are stored in a list-based data-structure. The simulation kernel then executes these SystemC processes by iterating through a list that contains all the ready-to-run SystemC processes. Therefore, hierarchical decomposition of a model during simulation is completely flattened. This impedes us from nesting DE component within other domains such as FSM and SDF. It limits us to having only one simulation context for DE and SystemC's DE kernel is essentially the master kernel rendering such composition of multiple DE components invalid so far.

An alternative for allowing HBH in SystemC for the DE MoC is to implement a secondary DE kernel. We plan to reuse several SystemC data-structures, data-types and channels as well as the threading packages, but change the design to allow nesting of DE inside other MoCs and exploring temporal synchronization of different DE components. We reserve a detailed discussion on this for another paper.

4.2 The Shallow: Mimicking it by Language Constructs Only

Language-level support for MoCs means that there are constructs in the language of the SLDL that describe a model following the desired MoC. For example, SystemC follows a DE-based MoC and its process types, channels, etc. follow the evaluate-update paradigm, thus defining the language of SystemC to a DE MoC. In fact, other MoCs can be modeled using SystemC by extending primitive channels and using hierarchical channels [8, 10]. We present an example of the dining philosopher in Section 5.1, which is a CSP model, but implemented using specialized classes on top of SystemC-2.0.1 that simulates via the DE kernel. Therefore, one possible avenue for heterogeneity in SystemC is to create a library of channels and constructs that allow the user to describe models following a particular MoC. In [14, 10], CSP-specific channels are used for the dining philosopher problem serving as an example for using the underlying SystemC's DE kernel to simulate CSP-like models. We also provide example of the dining philosopher in Section 5 that shows CSP-specific channels constructed using language-level constructs of SystemC.

An SLDL that provides language-level support has the advantage that no simulation kernel extensions are required for the extended MoC and that the MoC-specific constructs are created using already existing constructs of the language. Especially with SystemC, language extensions through class library expressions does not even need a new compiler. On the downside, it is possible that the extended MoC may suffer in simulation performance when compared to an MoC-specific kernel. Even though providing kernel-level support for heterogeneity may result in increased simulation performance, there is obvious overhead of programming the extension.

5. EXAMPLES USING MOC EXTENSIONS AT KERNEL VERSUS LANGUAGE-LEVEL

We describe a solution to the dining philosopher problem that employs kernel-level MoC extensions for SystemC and compare it with a synonymous implementation using SystemC's reference implementation. This model uses the CSP and FSM kernel extensions. We provide code snippets to show how these models are constructed using the language-level additions to SystemC for the extensions. These snippets only show the essential constructs required to create the model and the remaining details of implementation is omitted for clarity and space.

5.1 Dining Philosopher Example using CSP and FSM MoCs

In [11], communicating sequential processes is introduced as a model of computation for concurrency that originally dates back to 1978 [11]. In this MoC, sequential processes are combined with process combinators to form a concurrent system of communicating components. The protocol for communication in such an MoC is fully synchronous as opposed to data flow networks. For example, in data flow networks, buffers in the channels connecting two computing entities are assumed, and, based on buffer size, the computations proceed asynchronous to each other, leaving the communicable data at the buffers for the other components to pick up as and when ready. Of course, in real implementations buffers are of limited size and hence often times requires process blocking. In

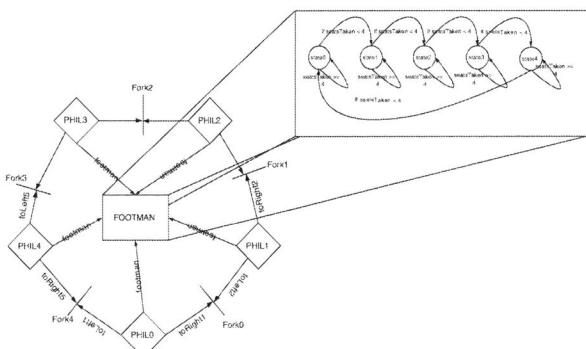

Figure 2: FSM and CSP Implementation of Dining Philosopher

CSP, the communication happens through a *rendez-vous* mechanism [11]. This necessitates synchronization at the data communication points between the processes, as buffering is not allowed on the channels, and the communicating processes both need to be ready to communicate for communication to take place. If one of the two communicating processes is not ready, the other blocks until both are ready.

To illustrate the CSP MoC, we take the classical *Dining Philosopher's* problem cited in [11]. The problem is defined as follows: there are five philosophers PHIL0, PHIL1, PHIL2, PHIL3, and PHIL4, there is one dining room in which all of them can eat, and each one has their own separate room in which they spend most of their time thinking. In the dining room there is a round table with five chairs assigned to each individual philosopher and five forks down on the table. In the middle of the table is a big spaghetti bowl that is replenished continuously (so there is no shortage of spaghetti).

A schematic of the way it can be implemented is shown in Figure 2. In addition, a deadlock may occur in the Dining Philosopher problem when for instance every philosopher feels hungry and picks up the fork to their left. That prevents any of the philosophers eating since two forks are required to eat. We use a simple deadlock avoidance technique where we implement a footman that takes the philosophers to their respective seats and, if there are four philosophers at the table, the footman asks the fifth philosopher to wait and seats him only after one is done eating. The CSP MoC models the philosophers and the forks, and the internals of the footman are governed by the FSM MoC.

Our implementation models the fork and philosopher entities as CSP processes. Declaration of the FORK module using DE and using CSP are shown in Listing 1 and Listing 2 respectively. Notice in Listing 1 the ports are instantiated with a template argument of sc_csp_channel_ifs, which were specifically constructed to model the rendez-vous semantics using the DE kernel. Using our extensions, the user simply uses the CSPport CSP-specific channel to allow for rendez-vous communication. However, the structure used in declaring a CSP process uses the SC_CSP_MODULE() macro that defines the CSP process, SC_CSP_CTOR() is the constructor and the SC_CSP_THREAD() initializes a CSP process for this module. The SC_CSP_THREAD() takes two arguments where the first is the entry function and the second informs the kernel of the CSP graph into which this CSP process is added. Declaration of the philosopher process is done in a similar fashion, but we do not present it in this paper for the sake of space.

Now, we present the implementation of one of the member func-

Listing 1: DE Fork Module

```
1 SC_MODULE(FORK) {
2   int id;  int queryFork;
3   int * drop; int * pick;
4   sc_port<sc_csp_channel_ifs<int>>
        fromRight;
5   sc_port<sc_csp_channel_ifs<int>>
        fromLeft;
6   void reqFork();
7   void addressFork();
8   SC_CTOR(FORK) {
9     queryFork = 1;
10    SC_THREAD(addressFork) {
11      sensitive << fromRight << fromLeft;
12    }
13  };
14 };
```

Listing 2: CSP Fork Module

```
1 SC_CSP_MODULE(FORK) {
2   int id;  int queryFork;
3   CSPport<int> fromRight;
4   CSPport<int> fromLeft;
5   int * drop;  int * pick;
6   void reqFork();
7   void addressFork();
8   SC_CSP_CTOR(FORK) {
9     queryFork = 1;
10    SC_CSP_THREAD(addressFork, DP);
11  };
12 };
```

Listing 3: State entry functions for Footman FSM

```
1 void s::state0() {
2   if (seatsTaken < 4) {
3     ++seatsTaken;
4     seatAvailable[0] = true;
5     fromPhil0->push(*giveSeat, *this);
6     fsm_model->setState("toplevel.state.
          state1");
7   }
8 };
9 // ... additional state declarations
```

Listing 4: DE FORK's reqFork()

```
1 void FORK::reqFork() {
2   while(true) {
3     if (queryFork == 1) {
4       int val = 0;
5       queryFork = -1;
6       forks[id] = queryFork;
7       fromLeft->put(*pick);
8
9       queryFork = 1;
10      forks[id] = queryFork;
11
12      fromLeft->get(val);
13
14      queryFork = -1;
15      forks[id] = queryFork;
16      fromRight->put(*pick);
17
18      queryFork = 1;
19      forks[id] = queryFork;
20      fromLeft->get(val);
21
22    } else {
23      if (queryFork == -1) {
24    int val = 0;
25    fromLeft->get(val);
26    if (val == 1) {
27      queryFork = 1;
28      forks[id] = queryFork;
29    }
30    else {
31      fromRight->get(val);
32      if (val == 1) {
33        queryFork = 1;
34        forks[id] = queryFork;
35      }
36    }
37    }
38    }
39  }
40 };
```

Listing 5: CSP FORK's reqFork()

```
1 void FORK::reqFork() {
2   while(true) {
3     if (queryFork == 1) {
4       queryFork = -1;
5       forks[id] = queryFork;
6       fromLeft.push(*pick, *this);
7
8       queryFork = 1;
9       forks[id] = queryFork;
10      fromLeft.get(*this);
11
12      queryFork = -1;
13      forks[id] = queryFork;
14      fromRight.push(*pick, *this);
15
16      queryFork = 1;
17      forks[id] = queryFork;
18      fromRight.get(*this);
19    } else {
20      if (queryFork == -1) {
21
22    int val = fromLeft.get(*this);
23    if (val == 1) {
24      queryFork = 1;
25      forks[id] = queryFork;
26    }
27    else {
28      val = fromRight.get(*this);
29      if (val == 1) {
30        queryFork = 1;
31        forks[id] = queryFork;
32      }
33    }
34    }
35  }
36  }
37 };
```

Figure 3: Code snippets for Fork's reqFork() implementation

tions of the FORK CSP process in Listing 4 and Listing 5. The implementations for the DE and CSP are similar for the reqFork() member function in that the implementation logic is the same except for the invocations to the channels. The main difference is in the arguments passed into the invocation. The CSP channels require passing the object from which the member function is invoked, hence the *this parameter. For the push() member function, we require the user to also pass the value to be put onto the CSP channel. The remainder of the implementation for the FORK and the PHIL modules are also similar and available on our website [5].

The role of the footman is to seat the philosophers to their designated seats and to monitor that only four philosophers are sitting at the dining table at any time. The implementation of the footman was done via a global function to have immediate verification of the state of the seats occupied at the dining table. However, to show the heterogeneity possible with our extension we take this example further by implementing the footman as an FSM embedded in a CSP process. For its corresponding DE design, we leave the implementation as a global function that checks the number of seats available partly because state machines are relatively simple to design using SystemC.

Figure 2 presents a state machine diagram showing the functions of the footman. The initial state is state0. The functionality of every state is the same except for the next state transitions. Every state has a self-loop suggesting that the control in the FSM does not transition to another state whenever four philosophers are

seated. However, if there are seats available, then a seat is allocated and the transition to the next state occurs. This is a simple FSM that changes the solution of the Dining Philosopher such that it ensures that every philosopher gets a turn to eat as well as serving as a deadlock avoidance mechanism. The module definition is shown in Listing 6 where object s is the state machine. The variables with prefix fromPhil are the ports through which the philosophers communicate with the footman requiring the footman to be encapsulated in a CSP process. Therefore, the FSM defining the behavior of the footman is embedded in a CSP process through which the CSP channels are tunneled. The implementation of the state entry functions are shown in Listing 3. Every state has identical implementation except for the next state transition.

The seatAvailable array maintains which seat has been occupied and a record of every philosopher to his particular seat is kept by the index of the array. For example, *seatAvailable[1]* refers to the seat that belongs to a philosopher with *id* one. The toplevel CSP process contained in fsmtop module is defined in Listing 7. This module definition has instances of *CSPport*s that have pointer declarations in Listing 6. The constructor of module fsmtop initialize an instance of s and appropriately assigns the addresses of the ports in object s1 to allow tunneling of the CSPports.

The main() function implementation for the DE and CSP implementations of the dining philosopher problem are shown in Listing 8 and Listing 9. The main() function for the DE model is straightforward SystemC with the only difference that we use our specific CSP channels sc_csp_channel. The CSP model's

Listing 6: Module definition Footman FSM

```
1 SC_FSM_MODULE(s) {
2   int random;
3   int * giveSeat;
4   CSPport<int> * fromPhil0;
5   CSPport<int> * fromPhil1;
6   CSPport<int> * fromPhil2;
7   CSPport<int> * fromPhil3;
8   CSPport<int> * fromPhil4;
9   void state0();
10  void state1();
11  void state2();
12  void state3();
13  void state4();
14  SC_FSM_CTOR(s) {
15    giveSeat = new int();
16    *giveSeat =1;
17    fsm_model->setState("toplevel.state.state0");
18    SC_FSM_METHOD(state0, fsm_model);
19    SC_FSM_METHOD(state1, fsm_model);
20    SC_FSM_METHOD(state2, fsm_model);
21    SC_FSM_METHOD(state3, fsm_model);
22    SC_FSM_METHOD(state4, fsm_model);
23  };
24 };
```

Listing 7: Toplevel CSP process for Footman

```
1 SC_CSP_MODULE(fsmtop) {
2   s * s1;
3   CSPport<int> fromPhil0;
4   CSPport<int> fromPhil1;
5   CSPport<int> fromPhil2;
6   CSPport<int> fromPhil3;
7   CSPport<int> fromPhil4;
8   void entry();
9   SC_CSP_CTOR(fsmtop) {
10    s1 = new s("state");
11    s1->csp = *this;
12    s1->fromPhil0 = &fromPhil0;
13    s1->fromPhil1 = &fromPhil1;
14    s1->fromPhil2 = &fromPhil2;
15    s1->fromPhil3 = &fromPhil3;
16    s1->fromPhil4 = &fromPhil4;
17    SC_CSP_THREAD(entry,DP);
18  };
19 };
```

Figure 4: Code snippets for Footman FSM and Toplevel CSP process for encapsulation

Listing 8: DE's main()

```
1 int main(int argc, char *argv[]) {
2   sc_clock clk;
3   sc_csp_channel<int> toRight1, toLeft2, toLeft1, toRight2;
4   // ... more instantiations of channels
5   FORK fk1("Fork_1");
6   fk1.fromLeft(toRight1);
7   fk1.fromRight(toLeft2);
8   fk1.drop = &drop;
9   fk1.pick = &pick;
10  fk1.id = 0;
11  // ... more instantiation of FORK
12  PHIL Philosopher1("PHIL_1");
13  Philosopher1.timeToLive = duration;
14  Philosopher1.drop = &drop;
15  Philosopher1.pick= &pick;
16  Philosopher1.toRight(toRight1);
17  Philosopher1.toLeft(toLeft1);
18  Philosopher1.id = 0;
19  // .. more instantiations of PHILs
20  sc_start(-1);
21  return 0;
22 };
```

Listing 9: CSP's main()

```
1 int main(int argc, char *argv[]) {
2   CSPchannel<int> toRight1, toLeft2, toLeft1, toRight2;
3   // ... more instantiations of channels
4   FORK fk1("Fork_1");
5   fk1.setprocname("FORK_1");
6   fk1.fromLeft(toRight1);
7   fk1.fromRight(toLeft2);
8   // ... more instantiations of FORKs
9   PHIL Philosopher1("PHIL_1");
10  Philosopher1.toRight(toRight1);
11  Philosopher1.toLeft(toLeft1);
12  Philosopher1.id = 0;
13  Philosopher1.footman(getSeat0);
14  // ... more instantiations of PHILs
15  fsm_model = new FSMReceiver("fsm1");
16  fsm_kernel.insert(fsm_model);
17  fsmtop myfsm("toplevel");
18  myfsm.fromPhil0(getSeat0);
19  myfsm.fromPhil1(getSeat1);
20  myfsm.fromPhil2(getSeat2);
21  myfsm.fromPhil3(getSeat3);
22  myfsm.fromPhil4(getSeat4);
23  Philosopher1.points_to(fk1, toRight1);
24  Philosopher1.points_to(fk5, toLeft1);
25  Philosopher1.points_to(myfsm, getSeat0);
26  Philosopher2.points_to(fk1, toLeft2);
27  Philosopher2.points_to(fk2, toRight2);
28  Philosopher2.points_to(myfsm, getSeat1);
29  // ... more connections
30  sc_csp_start("0",&DP);
31  return 0;
32 };
```

Figure 5: Code snippets for Dining Philosopher main()

main() is more involved starting by first instantiating the CSP specific channels. Then one instance of the FORK CSP process is shown. This is where the channels are bound to the ports and variables assigned. Similarly done for the PHIL object. A difference between the two is of the footman, which was mentioned earlier. We employ the FSM MoC to model the footman that also requires initialization and binding. Finally the CSP graph is created using the points_to() member function following Figure 2. In essence, this creates the graph informing the CSP kernel which CSP processes are connected to which other processes. Detailed explanation of the actual implementations and line by line description is available at [14].

In this example, the DE model seems synonymous to the CSP model aside from the footman implementation. This was possible by implementing a primitive SystemC channel such that the DE model hides the details of the rendez-vous communication within the channel. The CSP kernel's implementation also essentially uses the idea of a process informing the kernel that it has sent data, which invokes the kernel to respond by only transferring the data once the receiving process is ready to receive. The construction of the CSP processes and the DE modules are once again very similar and with the help of the CSP-specific primitive channel, the put() and get() are similarly invoked in both models. The difference between the two models is the construction of the CSP graph and the need for CSP-specific primitive channel. We omit the channel implementation for sake of space but the full implementation is available in [14, 5].

6. SIMULATION RESULTS

To investigate the performance of simulation models based on the SDF, CSP and FSM extensions to their counterparts in the DE kernel, we display results for the FIR as our SDF example, Dining Philosopher and Producer/Consumer as our CSP examples and a traffic light controller as our example for the FSM kernel. The Producer/Consumer and the traffic light controller examples are additional examples that we implement for experimentation.

Figure 5.1 shows that the SDF model of the FIR shows approximately 75% improvement over its DE counterpart. The SDF model

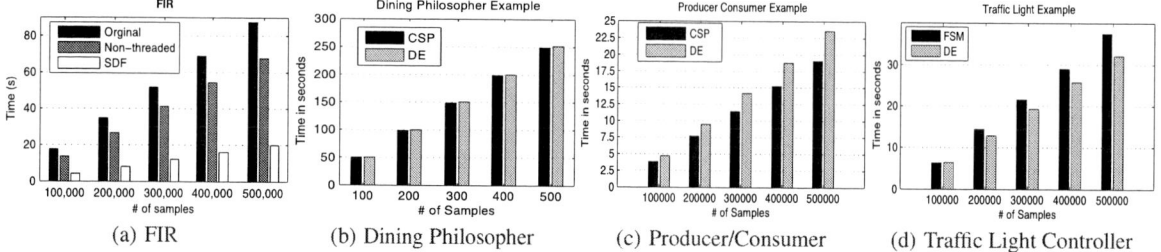

(a) FIR   (b) Dining Philosopher   (c) Producer/Consumer   (d) Traffic Light Controller

Figure 6: Results from Experiments

takes advantage of static scheduling of the SDF MoC whereas the DE model must follow its dynamic scheduling based on the evaluate-update paradigm. For the the CSP kernel, we see that the Producer/Consumer example performed better than its DE counterpart by approximately 18%, whereas the Dining Philosopher example showed to be only 1% better than the DE version using CSP primitive channels. The Producer/Consumer example shows a better result for the CSP kernel primarily due to the fact that there are only two CSP processes and there is limited synchronization within the implementation of the processes. The Dining Philosopher example is comparatively a more involved example with ten CSP processes requiring a significantly increased amount of synchronization. The worst case we believe is that the CSP kernel will perform as poorly as the DE kernel, but in most cases it will perform slightly better. The FSM kernel, on the contrary, performs approximately 10% slower than the implementation in a DE kernel. This drastic difference is understood by the implementation of the DE model, where one large `switch(...)` statement is used to model the FSM. This is very different from using the FSM kernel, where every `case` from the switch statement is a `SC_METHOD()` process incurring the overhead.

7. DISCUSSIONS

Determining whether shallow or deep heterogeneity is a better strategy for CAD mandates looking at the following aspects of modeling, simulation and capabilities of the SLDL:

- modeling ease in describing design.
- simulation efficiency of described design.
- possibilities for static analysis, deadlock checking, and verification.
- automated synthesis.

Even though modeling ease is a subjective measure, it is important to identify the difficulties an average SystemC user encounters when using the extensions for SystemC. From our examples, it is evident that the module construction in SystemC is very similar to the module declarations for the extensions as well. Slight changes such as arguments to be passed in the MACROs is expected, hence a regular SystemC user would have little problem in implementing module declarations using the extended kernels provided there is appropriate documentation. A user that understands the MoC being modeled and the modeling guidelines will encounter little problems in describing the implementation of the modules since the implementations are simpler than those of its counterparts using the DE MoC. However, all the extended kernels require a graph description of the model, which is not needed in SystemC DE models and thus adding an additional step to the modeling. However, SystemC's DE models require certain synchronizations embedded

within the implementations of the modules, such as handshaking, waits and rendez-vous communication. These synchronizations are not required when using the MoC extensions, because the MoCs themselves dictate the synchronization of the processes. In addition, if a user's design needs an MoC such as the CSP MoC, then specific CSP channels must be written, which requires a higher level of understanding of SystemC and programming skills. Having mentioned that, with the examples and documents available at the SystemC web-site [20], learning how to program hierarchical and primitive channels may not be that difficult of a task. Hence, in terms of modeling ease, we find that shallow and deep are equally poised as good approaches. This is primarily because both approaches require a certain level of understanding and learning.

Implementing the execution semantics for specific MoCs in the extended kernels promotes the idea that simulation is faster. However, our experiment results do not conclusively show that for every MoC the simulation is faster. The SDF MoC shows significant gains, whereas the FSM and CSP extensions lack in their simulation performance. We attribute the increase in SDF models to being able to statically schedule the execution of the model during initialization. This rids the need for notifying and updating events as done when using the SystemC's DE MoC. We attribute the poor performance exhibited by the FSM and CSP MoCs to the implementation specifics of the kernels. For example, the FSM MoC uses a one-process-per-state implementation, which is not a good design decision because every FSM state suffers from context-switching overhead due to its underlying `SC_METHOD` implementation. A better implementation of the FSM MoC is available at [15], which reports no performance degradation in comparison to a `switch/case` implementation of an FSM, which is one of the fastest way to implement an FSM using C/C++. Similarly, the CSP MoC suffers due to implementation decisions. By improving the implementation of the MoCs, it is possible to at least maintain the same simulation times as that of SystemC's DE MoC with the additional capabilities of HBH. From our simulation experiments we find that deep heterogeneity can be used to create designs such that it is better or equivalent in terms of simulation efficiency.

Enforcing MoC-based design via kernel-level extensions helps in performing static analysis and checking whether a design reaches any unsafe state, such as a deadlock state. These are currently not available in SystemC-H, but it is possible to analyze the graphs for certain characteristics. Another interesting possible use of MoCs at the kernel-level is for behavioral synthesis, which is considered a difficult problem. Having the user describe models using MoCs provides more information to the kernels for simulation, but it also provides more model information for synthesis back-ends. For example, Ptolemy II [4] effectively utilizes the MoC descriptions to generate Java code. Similarly, it is possible to use MoCs for constraining the behaviors described in a model and easing the task of behavioral synthesis. So, for analysis and synthesis capabilities, it is better to have specific kernel-level MoCs.

It is important to note that SystemC is a special SLDL because of its open-source nature. Industry compilers for other hardware description languages do not allow such freedom. This makes it difficult to extend such languages natively. However, it may be beneficial for such tools to consider allowing interfaces to override the existing simulation kernel in the form of a plug-in. So, in terms of flexibility of most SLDLs (aside from SystemC) it is better to employ shallow heterogeneity since most SLDLs do not provide an infrastructure for easily extending the simulation kernels.

Finally, SystemC-AMS [21, 7] is an interesting case because it provides an MoC for continuous-time and analog mixed signal modeling and simulation. Such an MoC is hard to describe using SystemC because it requires differential equation solvers and specific methods for simulating analog components. Even if SystemC was to have solver libraries, the synchronization requirements for cleanly interacting with regular DE semantics would be a much arduous task than one presented by the SystemC-AMS group. Based on this brief presentation of the issues with deep versus shallow heterogeneity, we claim that deep heterogeneity is the better of the two approaches for improved capabilities for analysis, synthesis and fast simulation.

8. FUTURE WORK & CONCLUSION

The successful implementations of SDF in [16], CSP and FSM kernel extensions for SystemC thoroughly documented in [14], provide a motivation towards heterogeneity in SystemC. Taking these advancements in heterogeneous SystemC, a solid foundation for supporting heterogeneous behavioral hierarchy can be envisioned. Furthermore, as a supplementary objective, the possibility of improving simulation efficiency by taking advantage of specific MoC behaviors is essential with the large sizes and complexities off designs. However, our experimentation reveals that in terms of modeling ease, shallow nor deep heterogeneity give significant leverage. Additionally, the simulation experiments disallow us from conclusively stating that one strategy is better than the other suggesting that perhaps the kernel implementations require refinements for fast simulation. The opportunity for static analysis, verification and automated synthesis are all very attractive characteristics of deep heterogeneity, however, extensions for SystemC as of now do not support such techniques. Nonetheless, the potential in an environment that supports deep heterogeneity outweighs one which employs shallow heterogeneity especially in terms of analysis and we claim that deep heterogeneity has better prospect for CAD industries.

9. REFERENCES

[1] S. Bhattacharyya, P. Murthy, and E. Lee. *Software Synthesis from Dataflow Graphs*. Kluwer Academic Publishers, 1996.

[2] D. A. Mathaikutty. Functional programming and metamodeling frameworks for system design. Master's thesis, Virginia Tech, Blacksburg, Virginia, May 2005.

[3] E. A. de Kock, G. Essink, W. J. M. Smits, P. van der Wolf, J. Y. Brunel, W. M. Kruijtzer, P. Lieverse and K. A. Vissers. YAPI: Application Modeling for Signal Processing Systems. In *the proceedings of Design Automation Conference*, 2000.

[4] E. Lee et al. Heterogenous Concurrent Modelling and Design in Java: Volume 2 - Ptolemy II Software Architecture. Memorandum UCB/ERL M03/28, July 2003.

[5] FERMAT. Formal Engineering Research with Models, Abstractions and Transformations. http://fermat.ece.vt.edu.

[6] F. Ghenassia, editor. *Transaction-level Modeling with SystemC*. Springer, 2005.

[7] C. Grimm. Modeling and Refinement of Mixed Signal Systems with SystemC. In *Methodologies and Applications. Kluwer Academic Publisher*, 2003.

[8] T. Grotker, S. Liao, G. Martin, and S. Swan. *System Design with SystemC*. Kluwer Academic Publishers, 2002.

[9] Metropolis Group. Metropolis: Design environment for heterogeneous systems. Website: http://embedded.eecs.berkeley.edu/metropolis/index.html.

[10] F. Herrera and E. Villar. A Framework for Embedded System Specification under Different Models of Computaiton in SystemC. In *Design Automation Conference*, 2006.

[11] C. A. R. Hoare. *Communicating Sequential Processes*. Prentice Hall, 1985.

[12] E. A. Lee and D. G. Messerschmitt. Static Scheduling of Synchronous Data Flow Programs for Digital Signal Processing. In *the Proceedings of IEEE Transactions on Computers*, volume Vol. C-36 of *NO. 1*, 1987.

[13] Edward A. Lee and Alberto L. Sangiovanni-Vincentelli. Comparing Models of Computation. In *the Proceedings of the International Conference on Computer-Aided Design (ICCAD)*, pages 234–241. IEEE Computer Society, 1996.

[14] H. D. Patel and S. K. Shukla. *SystemC Kernel Extensions for Heterogeneous System Modeling*. Kluwer Academic Publishers, 2004.

[15] H. D. Patel and S. K. Shukla. Heterogeneous Behavioral Hierarchy for System Level Designs. In *the Proceedings of Design Automation and Test in Europe*, 2005.

[16] H. D. Patel and S. K. Shukla. Towards a heterogeneous simulation kernel for system level models: A systemc kernel for synchronous data flow models. In *IEEE Transactions in Computer-Aided Design*, volume 24, August 2005.

[17] H. D. Patel and S. K. Shukla. Towards behavioral hierarchy extensions for systemc. In *Forum on Design and Specification Languages (FDL '05)*, 2005.

[18] Ptolemy Group. Ptolemy II Website. http://ptolemy.eecs.berkeley.edu/ptolemyII/.

[19] SML-Sys Framework. SML-Sys Website. http://fermat.ece.vt.edu/SMLFramework, 2004.

[20] SYSTEMC. SystemC. Website: http://www.systemc.org.

[21] A. Vachoux, C. Grimm, and K. Einwich. SystemC Extensions for Heterogeneous and Mixed Discrete/Continuous Systems. In *International Symposium on Circuits and Systems*, 2005.

Statistical Static Timing Analysis Considering the Impact of Power Supply Noise in VLSI Circuits

Hyun Sung Kim, D. M. H. Walker

Dept. of Computer Science
Texas A&M University
College Station TX 77843-3112
Tel: (979) 862-4387
Fax: (979) 847-8578
Email: {hskim, walker}@cs.tamu.edu

Abstract

As semiconductor technology is scaled and voltage level is reduced, the impact of power supply variation has become very significant in predicting the realistic worst-case delays in integrated circuits. The analysis of power supply noise is inevitable since there is a high correlation between delay and supply voltage. Supply noise analysis has often used a vector-based STA approach. However, vector-based approaches are very expensive, particularly during the design phase. In this work, two novel vectorless approaches are described such that circuit delay increases due to power supply noise can be efficiently estimated. Experimental results on ISCAS89 circuits show not only the accuracy of our approaches, but also the indispensability of considering care-bits, which sensitize the longest paths during the power supply noise analysis.

1. Introduction

In deep submicron (DSM) technology, power supply analysis has become increasingly important in predicting the realistic worst-case delays in integrated circuits [1]. Fluctuations of 10% in power/ground supply voltage can cause the delay for standard gates to vary by up to 30% in 130 nm technology [2][3]. Newer technologies have increased delay sensitivity to supply noise, due to reduced gate overdrive. In addition to causing reduced circuit performance, supply noise can cause functional failure. Therefore, the analysis of power supply noise has become inevitable in timing analysis.

Generally, the supply voltage noise is due to both the parasitic resistance (IR) and inductance $(L \cdot di/dt)$ of on-chip and package interconnect. The on-chip power grid is predominately resistive, with its noise produced by IR drop. Package interconnect has a higher inductance, so its noise is generated primarily by $L \cdot di/dt$ effects. At faster gate switching speeds and higher circuit density, on-chip inductance must be taken into account [3].

Power supply voltage analysis has been addressed through vector-based and vectorless approaches. Many vector-based approaches [1][4][5][6][7] use genetic or other algorithms to find a set of input vectors which cause the maximum voltage drop on certain targeted regions, whereas the vectorless approaches [2][8][9] use circuit

timing and functional information, a superposition method and supply current constraints. Previously, supply noise analysis has often used logic simulation. However, in future DSM technology, the vectorless technique is expected to be favored due to the cost of simulation.

We propose two novel vectorless approaches to incorporating supply voltage noise analysis into static timing analysis (STA). These approaches use a set of vectors produced by input pattern generation methods to statistically estimate the realistic power supply noise. We use a supply noise modeling approach developed for delay test generation [4]. Due to the correlation of supply voltage noise between circuit blocks, we adopt the Principal Component Analysis (PCA) technique [11]. The PCA technique not only identifies a small set of uncorrelated parameters that explain most of the noise for the circuit blocks, but it also transforms the set of correlated parameters into a set of uncorrelated parameters. Once we obtain all uncorrelated supply voltage variation across the chip, we can determine delay distributions corresponding to the supply voltage distributions for all individual gates on the chip by using a linear delay model [10] and the sensitivity of delay to supply voltage. Then, we can perform statistical static timing analysis by propagating delay distributions along the longest paths on the chip, and finding the maximum of these distributions. We avoid the abbreviation SSTA, since it is usually associated with statistical process variations, rather than statistical supply voltage variations.

This paper is organized as following. In section 2, we introduce a *static* technique to estimate the supply voltage noise distribution across all regions on the chip. We then perform statistical static timing analysis with the supply voltage noise distribution in section 3. Section 4 shows experimental results of our approach on ISCAS89 benchmark circuits. Finally, a conclusion is provided in section 5.

2. Estimation of Voltage Noise Distributions

A fast, accurate estimation of the power supply noise distribution is essential for accurate timing analysis. In this section, we present a vectorless approach to estimate the power supply noise distribution based on both the

---

1550-4093/07 $25.00 © 2007 IEEE

simplified power region model and the circuit switching model proposed in [4].

2.1. Power Region and Switching Models

Because RLC network analysis is expensive, a simplified power region has been proposed. The maximum voltage drop ΔV_{max} in a region during a clock cycle can be estimated with several approximations described in [4]:

$$\Delta V_{max} = \left(\sum_{i=0}^{n} \int I_{sc\_i} \right) / \left(C_d + C_p \right) \qquad \text{(EQ 1)}$$

where C_d and C_p are respectively the single lumped decoupling capacitance and the total parasitic capacitances of devices and interconnect connected to the power supply network in a region. The switching current, denoted as $\sum_{i=0}^{n} \int I_{sc\_i}$, is the summation of currents that flow into all n switching gates in the region during the clock cycle.

We also employ the circuit switching model, which is similar to the approximated models proposed in [3][10], to estimate the switching current. The switching current consists of leakage current and charging/discharging current. We do not discuss leakage current further, since we treat it as constant, and analyze its voltage impact with a one-time IR drop analysis.

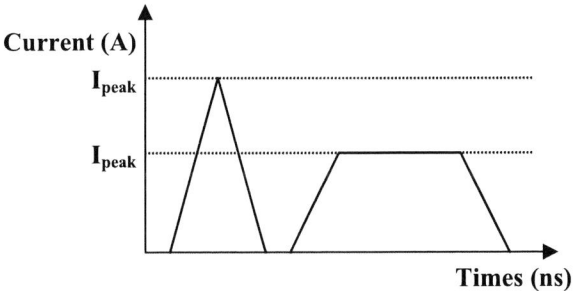

Figure 1. Triangular and trapezoidal current waveforms.

As shown in **Figure 1**, switching current is approximated by a piecewise linear current waveform, which is triangular for small load capacitances and trapezoidal for large load capacitances.

2.2. Statistical Model for Supply Voltage Noise

The chip is divided into a rectangular grid, with gates assigned to the regions where they connect to the power grid. We run zero-delay logic simulation on the circuit with three different types of input patterns: MC (Monte Carlo approach), AAC (analytical approach with care bits), and AAR (analytical approach with random bits). We first generate test patterns propagating transitions along the top

200 longest paths in the circuit using the *CodGen* ATPG tool [14]. In the MC approach, all "don't care" bits in the test patterns are randomly filled. In the AAR approach, for the purposes of noise analysis, all input bits are random, including the care bits. In the AAC approach, a set of care bits are selected as follows: we choose the path, among the 200 longest paths, which had the highest probability of being the critical path among 1000 MC samples. The "don't care" bits are randomly filled.

The input patterns for MC, AAR, and AAC are simulated to obtain statistical parameters of supply voltage noise distributions for each region. We assume that the random variables for each region are Normal. Because of correlations in voltage noise distributions between regions, we employ the PCA technique. The PCA method transforms a set of correlated random variables $\overline{X} = \{x_1, x_2, ..., x_n\}$ with a covariance matrix M into a set of uncorrelated random variables $\overline{X'} = \{x'_1, x'_2, ..., x'_n\}$, such that any random variable $x_i \in \overline{X}$ can be expressed as a linear function of the principal components with 0 mean and 1 variance in $\overline{X'}$:

$$x_i = \mu_i + \left(\sum_j \sqrt{\lambda_j} \cdot v_{ij} \cdot x'_j \right) \cdot \sigma_i \qquad \text{(EQ 2)}$$

where μ_i and σ_i are the mean and the standard deviation of x_i, λ_j is the jth eigenvalue of the covariance matrix M, v_{ij} is the ith element of the jth eigenvector of M and $x_j \in \overline{X'}$ [13]. The PCA technique is incorporated in both statistical voltage noise analysis and statistical timing analysis for quick, efficient computation.

3. STA with Power Supply Noise Variation

With the statistical parameters from the fast power supply noise analysis, we can statistically evaluate the performance of the circuit. Here, we consider temporal and spatial supply voltage noise variation.

3.1. Temporal and Spatial Voltage Variation

Because power supply noise and logic gate switching times are both uncertain, it is very difficult to determine the supply voltage at the time a logic gate switches. We adopt the approximation proposed by Wang [4], using the average of the initial and worst-case supply voltages during the clock cycle.

In addition to temporal variation, supply voltage has spatial variation. If driver and receiver gates are far enough apart, they can have different supply voltages. This can significantly affect the gate delay because the charging/discharging current heavily depends on the input supply voltage. Hashimoto [12] proposed PG level equalization – after equalizing input supply voltage and gate supply voltage, the output load capacitance is increased/decreased by the same ratio. However, we found

1550-4093/07 $25.00 © 2007 IEEE

that Hashimoto's method does not work well over our range of output loads and input slopes. We obtained more accurate results by equalizing the input and gate supply voltage without changing the output load capacitance.

3.2. Gate Delay Model and Path Computation

We employ the gate delay model proposed in [12] to calculate the gate delay and output transition time:

$$t_d = f\left(t_{in}, C_{load}, V_{receiver}\right) \qquad \text{(EQ 3)}$$

$$t_{out} = g\left(t_{in}, C_{load}, V_{receiver}\right) \qquad \text{(EQ 4)}$$

where t_{in} and t_{out} are the input and output transition time, respectively, t_d is the gate delay, C_{load} is the output load capacitance and finally $V_{receiver}$ is the receiver supply voltage.

The supply voltage is not a deterministic value, but a random variable. We utilize the sensitivity of supply voltage versus delay to compute an individual gate delay distribution.

$$\delta = \frac{\left(f\left(t_{in}, C_{load}, V_{\mu}\right) - f\left(t_{in}, C_{load}, V_{\mu \mid \sigma}\right)\right)}{\left(V_{\mu+\sigma} - V_{\mu}\right)} \qquad \text{(EQ 5)}$$

where V_{μ} and $V_{\mu+\sigma}$ are the mean and (mean + standard deviation) of a random variable V, respectively, and δ is the sensitivity of delay versus voltage. Once individual gate delay random variables are computed, the longest path computation as well as circuit performance analysis with the longest paths can be easily done using the *sum* and *max* functions of PCA properties described in [11].

4. Experimental Results

In this experiment, ISCAS89 benchmark circuits are implemented in 180 nm with 1.8 V static CMOS technology. We use the *CodGen* ATPG tool [14] to automatically generate a set of longest paths and corresponding set of path-dependent input patterns. The input patterns consist of "don't care" bits and care bits, where the care bits sensitize the longest paths.

In our first experiment, we first validate the accuracy of the Monte Carlo approach which employs the simplified power grid and gate delay models, by comparing results with Cadence *Spectre* simulation, denoted as SS. For validation, we select the path with the highest probability of being the longest from the combinational version of ISCAS89 benchmark s1488. All "don't care" bits in the sensitizing pattern are randomly filled in our first experiment. A set of these randomly filled patterns is generated, and simulated, to obtain the simulated voltage drop distribution.

Figure 2 shows the distributions of supply voltage drops across circuit s1488 circuit using the MC and SS methods. Whereas the SS voltage drop distribution can be approximated as a Normal distribution, MC distribution cannot. One possible reason the MC distribution is not

Normal is the very small number of gates (673) in circuit s1488. Figure 3 shows that the MC distribution of voltage drops in the larger s38417 circuit is close to Normal. Although the means of the MC and SS voltage drop distributions are close, there is a large difference in their standard deviations. The simplified power region, approximated circuit switching models, and the small number of gates in s1488 are possible explanations for this difference.

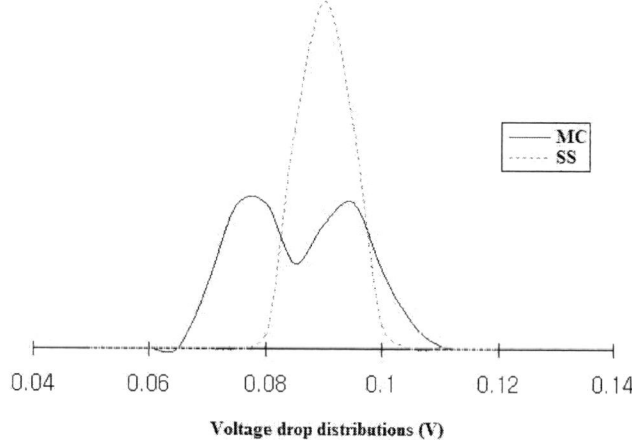

Figure 2. Voltage drop distributions of MC and SS in s1488.

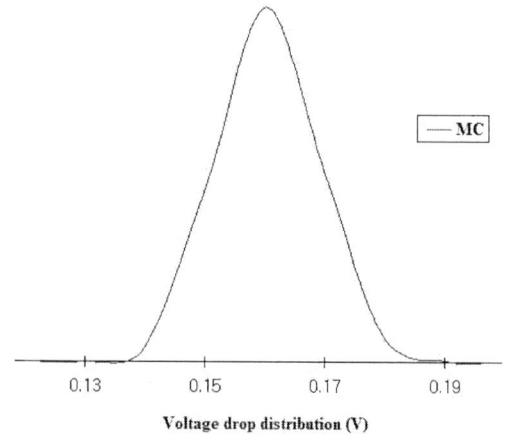

Figure 3. A voltage drop distribution in s38417.

As shown in Figure 4, we compute the realistic worst-case delay distribution of the longest path in s1488 circuit, using the voltage noise computed using MC and SS approaches. The differences between MC and SS worst-case delay distributions are much smaller than those between the voltage drop distributions in Figure 2. In other words, the differences in the standard deviations of voltage drop distributions in Figure 2 have little impact on the delay distributions. This may be because of the relatively low sensitivity between delay and supply voltage in the

180 nm technology, or due to an averaging effect. We will further investigate this in much larger benchmark circuits in the future. The results in Figure 2 and Figure 4 are summarized in Table 1.

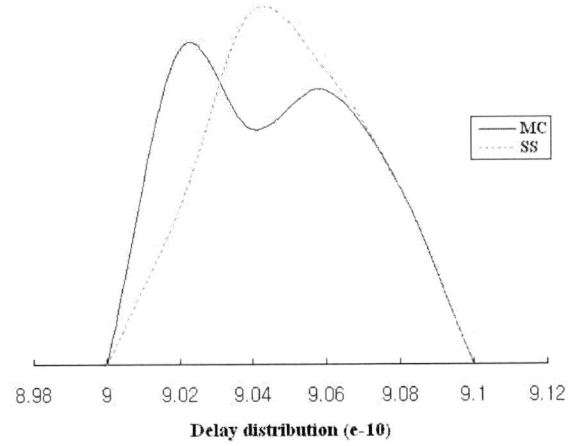

Figure 4. Path delays in MC and SS.

Table 1. Voltage drops and delays in s1488.

Circuit s1488	MC		SS		Error	
	Voltage drop (V)	Delay (ps)	Voltage drop (V)	Delay (ps)	Due to Voltage (%)	Due to Delay (%)
μ	0.083	904	0.088	904	5.68	0
σ	0.010	2.21	0.004	1.82	60	17.6

In our second experiment, we apply the MC calculation for a large number of input patterns, to compute the circuit delay distribution. Circuit simulation is too expensive to generate this large number of samples. We then compare these results to our proposed analytical approaches. Unlike the first experiment, we use the 200 longest paths in each circuit, and also group gates into regions. Here, we use 9 regions for each benchmark circuit.

The MC approach computes the *max* of the 200 path delay distributions numerically in the STA computation. This should provide higher accuracy than the analytical *max* approximations commonly used in statistical static timing analysis, but at higher cost. Since the MC approach is time-consuming, we developed two types of analytical approaches: with and without considering the care bits required to propagate transitions on the longest circuit paths, denoted as AAC and AAR, respectively.

As described in Section 2, AAC first performs a statistical voltage noise analysis with care bits. We then use PCA to transform the set of correlated voltage random variables across regions on the chip into a set of uncorrelated voltage random variables. Given the uncorrelated voltage random variables, we employ the gate delay model as well as the sensitivity model to compute

the gate delay distribution. Then, we propagate all gate delay distributions using the *sum* operation along the 200 longest paths. Finally, we obtain the circuit performance by applying an analytical *max* operation to all 200 path delay distributions. Since we do voltage noise analysis and timing analysis statistically, we only need to perform this analysis once for each benchmark circuit. That is why this approach is very fast when compared to MC. The AAR approach is identical to AAC, except that all input bits are random.

Table 2 shows the means and standard deviations of delay distributions calculated by the MC, AAC, and AAR approaches. The μ and σ from AAR for circuit s1488 are farther off from the MC values than those for circuits s35932 and s38417. As with the first experiment, this may be due to the small size of s1488. It can be seen that using random inputs rather than longest-path care bit values causes an underestimate in mean delay and overestimate in standard deviation.

We see a similar, but less severe phenomenon in s35932. There is little difference in mean delay, but AAC reduces the error in σ by 36 % compared to AAR. This is surprising considering that only 0.2% of the input bits are care bits. This small number of bits causes a noticeable change in supply noise and delay variation.

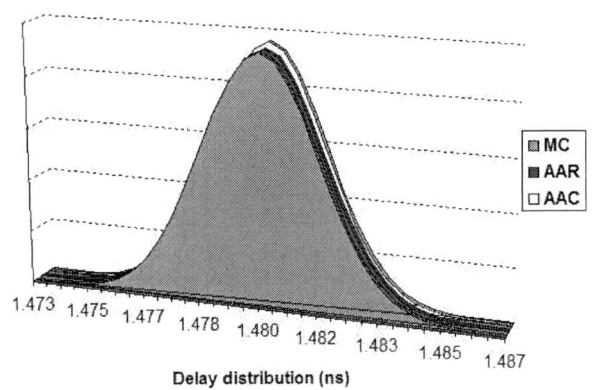

Figure 5. Delay distributions in s35932.

Unlike in s1488 and s35932, AAR results for the worst case delay μ and σ match that of the MC results in circuit s38417. Figure 5 illustrates the accuracy of the analytical AAC and AAR approaches versus the MC numerical approach in the s35932 benchmark circuit. One reason for the accuracy in s35932 may be that the larger number of gates causes the noise to appear more random, and the longer paths causes more averaging. The other reason may be that the care bits for the longest paths in this design have less impact on supply noise. Note that even though the noise-induced delay variation in Figure 5 is small, this is only due to the power grid design, and does not affect the accuracy of the analysis technique.

1550-4093/07 $25.00 © 2007 IEEE

Table 2. Statistical parameters of delay distributions in MC, AAR, AAC.

Benchmarks	MC			AAR			AAC		
	μ (ps)	σ (ps)	CPU Time	μ (ps)	σ (ps)	CPU Time	μ (ps)	σ (ps)	CPU Time
s1488	927	3.20	264 s	901	11.2	11 s	927	3.06	11 s
s35932	336	0.36	1.54 hr	335	1.03	107 s	336	0.26	107 s
s38417	1480	1.69	2.52 hr	1480	1.69	134 s	1480	1.66	134 s

In circuits s35932 and s38417, the data in Table 2 shows that the σ values for the AAC approach are still off from the σ values from the MC method. This is because in these circuits, there are multiple paths with significant probability of being the longest path for any one random pattern. The standard deviation errors can be reduced by intelligently deciding which care bits should be used in the analysis. Since AAR and AAC both take the same amount of CPU time, if the care bit information is available, then the AAC method is preferable to increase the accuracy of the σ calculation. The CodGen ATPG time for these circuits was only a few seconds, so obtaining care bits should usually not be a problem.

Finally, we observe that both AAR and AAC are much faster than the MC approach. Statistical power noise analysis and timing analysis reduce run time. Thus, the analytical approaches can be very helpful for quickly estimating the impact of supply noise during early design phases.

5. Conclusions

In deep submicron technology, power supply noise analysis must be performed during timing analysis. Supply noise analysis has often used a vector-based approach. However, this is very expensive, particularly during the early design phase. In this paper, we introduce novel vectorless approaches, with and without considering care bits, which sensitize the longest paths in the circuit. These methodologies can be used efficiently and accurately to estimate the delay increases due to power supply noise. Our experiments on ISCAS89 circuits also demonstrate the importance of a small number of care-bits during the power supply noise analysis.

Acknowledgements

We would like to thank Jing Wang for her thoughtful discussions about the simplified power region model.

References

[1] Y. M. Jiang and K. T. Cheng, "Analysis of Performance Impact Caused by Power Supply Noise in Deep Submicron Devices," *ACM/IEEE Design Automation Conf.*, New Orleans, LA, June 1999, pp. 760-765.

[2] S. Pant, D. Blaauw, V. Zolotov, S. Sundareswaran and R. Panda, "Vectorless Analysis of Supply Noise Induced Delay Variation," *IEEE/ACM Int'l Conf. Computer Aided Design*, San Jose, CA, Nov. 2003, pp. 184-191.

[3] H. H. Chen and D. D. Ling, "Power Supply Noise Analysis Methodology for Deep Submicron VLSI Chip Design," *ACM/IEEE Design Automation Conf.*, Anaheim, CA, June 1997, pp. 638-643.

[4] J. Wang, Z. Yue, X. Lu, W. Qiu, W. Shi and D. M. H. Walker, "A Vector-based Approach for Power Supply Noise Analysis in Test Compaction," *IEEE Int'l Test Conf.*, Austin, TX, Nov. 2005.

[5] Y. M. Jiang, A. Krstic, and K. T. Cheng, "Dynamic Timing Analysis Considering Power Supply Noise Effects," *Int'l Symp. on Quality of Electronic Design*, San Jose, CA, March 2000, pages 137-143.

[6] J. J. Liou, A. Krstic, Y. M. Jiang and K. T. Cheng, "Path Selection and Pattern Generation for Dynamic Timing Analysis Considering Power Supply Noise Effects," *IEEE/ACM Int'l Conf. Computer Aided Design*, San Jose, CA, Nov. 2000, pp. 493-497.

[7] A. Krstic and K. T. Cheng, "Vector Generation for Maximum Instantaneous Current Through Supply Lines for CMOS Circuits," *ACM/IEEE Design Automation Conf.*, Anaheim, CA, June 1997, pp. 383-388.

[8] G. Bai, S. Bodda and I. N. Hajj, "Static Timing Analysis Including Power Supply Noise Effect on Propagation Delay in VLSI Circuits," *ACM/IEEE Design Automation Conf.*, Las Vegas, NV, Jun. 2001, pp. 295-300.

[9] D. Kouroussis and F. N. Najm, "A Static Pattern-Independent Technique for Power Grid Voltage Integrity Verification," *ACM/IEEE Design Automation Conf.*, Anaheim, CA, June, 2003, pp. 99-104.

[10] C. Tirumurti, S. Kundu, S. Sur-Kolay and Y.-S. Chang, "A Modeling Approach for Addressing Power Supply Switching Noise Related Failures of Integrated Circuits," *Design, Automation and Test in Europe Conf. and Exhibition*, Paris, France, Feb. 2004, pp. 1078-1083.

[11] D. F. Morrison, "Multivariate Statistical Methods," *New York: McGraw-Hill*, 1976.

[12] M. Hashimoto, J. Yamaguchi and H. Onodera, "Timing Analysis Considering Spatial Power/Ground Level Variation," *IEEE/ACM Int'l Conf. Computer Aided Design*, San Jose, CA, Nov. 2004, pp. 814-820.

[13] H. Chang and S.S. Sapatnekar, "Statistical Timing Analysis Considering Spatial Correlations Using A Single Pert-like Traversal," *IEEE/ACM Int'l Conf. Computer Aided Design*, San Jose, CA, Nov. 2003, pp. 621-625.

[14] W. Qiu, J. Wang, D. M. H. Walker, D. Reddy, X. Lu, Z. Li, W. Shi and H. Balachandran, "K Longest Paths Per Gate (KLPG) Test Generation for Scan-Based Sequential Circuits," *IEEE Int'l Test Conf.*, Charlotte, NC, Oct. 2004, pp. 223-231.

SECTION 4:
DESIGN ERROR DEBUG &
DIAGNOSIS

Debug Support for Scalable System-on-Chip

Jianmin Zhang
School of Computer Science
National University of
Defense Technology
Changsha, China 410073
Telephone: +86-731-4575981
Fax: +86-731-4575981
Email: jmzhang@nudt.edu.cn

Ming Yan
School of Computer Science
National University of
Defense Technology
Changsha, China 410073
Telephone: +86-731-4575981
Fax: +86-731-4575981
Email: mingyan@nudt.edu.cn

Sikun Li
School of Computer Science
National University of
Defense Technology
Changsha, China 410073
Telephone: +86-731-4575981
Fax: +86-731-4575981
Email: lisikun@263.net.cn

Abstract— On-chip debug is an important technique to detect and locate the faults in the practical software applications. Scalability and reusability are the essential features of System-on-Chip (SoC). Therefore, the debug architecture should meet the requirement of those features. Furthermore, it is necessary for applications developers to communication with the SoC chip on-line. In this paper, we present the novel debug architecture to solve above problems. The debug architecture has been implemented in a typical SoC chip. The results of performance analysis show that the debug architecture has high performance at the cost of few resources and area.

I. INTRODUCTION

With the development of integrated circuit, a complete system can be fully integrated into SoC containing a set of high-complexity IP cores. Consequently, the functional behavior of the systems is determined by both hardware and software events. It is a trend to reduce the difficulty and time to market of software applications development. However, detecting and locating the faults in the software is tedious, time-consuming and error-prone. On-chip debug is an important technique to solve above problems. An effective debug methodology necessarily needs to provide the convenient environment to debug the errors in software. Such debug architecture should easily debug of problems in the "grey area" where the cause of a problem is not so clear. Furthermore, scalability and reusability are the primary features of System-on-Chip. These features entail the IP cores from the third parties to easily integrate into SoC. Therefore, the debug architecture should meet the demand of scalability and reusability. While IP cores are added or removed, such debug architecture need not change a lot to adapt to the new SoC. Moreover, it is necessary and convenient for applications developer to understand the status of the processor real time without interrupting the programs' execution and switching to the debug state. Thus on-line communication plays an important role in locating the bugs of software. It is an indispensable function of the debug architecture.

There have been a lot of contributions to research on debug architecture of the SoC chip and board [1], [2], [3], and [4]. They have provided good solutions for system-on-chip, even for the multi-core system. However they do not pay attention to the scalability and reusability of System-on-Chip. Existing works have very little concern regarding debug architecture for scalable SoC. Moreover, on-line communication is important for debug the faults in the software.

In this paper, we present the novel debug architecture to address those problems. Several special scan chains and a clock domain converter are introduced to ensure the scalability required by SoC. We design a module called a debug data exchange channel to effectively solve the problem of on-line communication. The debug architecture has been implemented in a typical SoC chip called EStar, which can represent a category of single-core SoC chips. EStar contains a 32-bits RISC processor core with 5 stages pipeline and many peripheral IP cores partially from the third parties. However it is in unsaturation, that is to say, it will integrate more IP cores. This debug architecture has good scalability and reusability, and does not necessarily modify a lot to support new IP cores integrated. We made a thorough performance analysis for all debug modules and basic debug operations. The results indicate that the debug architecture has high performance at the cost of few resources and area.

The paper is organized as follows. The next section surveys the novel debug architecture of SoC. Section 3 presents the debug support solution for the SoC's scalable and reusable features. Section 4 describes the solution of on-line communication for System-On-Chip. Section 5 shows the results of resource utilization and performance analysis. Finally, Section 6 concludes this paper and outlines the future research works.

II. DEBUG ARCHITECTURE

In this section, the overview of the debug architecture for SoC is introduced. Firstly, we provide the sketch of the debug architecture. Secondly, the structure and function of each debug component are described respectively in the following paragraphs.

A. Overview of Debug Architecture

The debug architecture for SoC is composed of seven components: a JTAG controller, four scan chains, a debug unit, a clock domain converter, a debug data exchange channel, a clock switch unit and a debug controller. Fig. 1 illustrates the debug architecture of SoC.

1550-4093/07 $25.00 © 2007 IEEE

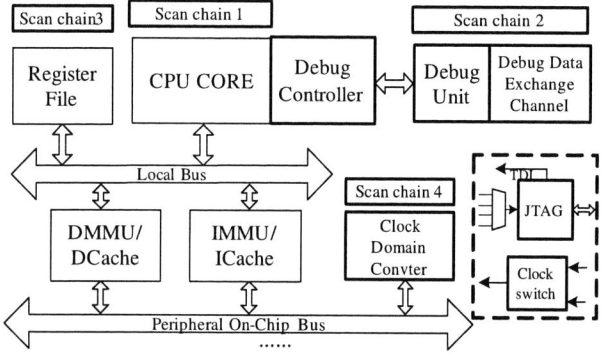

Fig. 1. Debug Architecture of SoC

B. Introduction of Debug Modules

The JTAG controller fulfills the communication between the debug target system and the environment. It receives debug instructions from the external debugger and translates them into the control signals of scan chains. It consists of a TAP (Test Access Port) controller, a debug instructions decoder and a boundary scan chain. It is fully compatible with IEEE 1149.1 standard Joint Test Access Group specification [5].

The debug scan chains are the bridges which connect the JTAG controller with the modules accessed. Data is serially shifted under the control of JTAG controller, but the scan chains access the modules in parallel. The debug architecture contains four debug scan chains. Scan chain 1 controls the instructions and data port of the processor core. Scan chain 2 sets the address and data registers of the breakpoints and watchpoints in debug unit, and transmits message between JTAG controller and the debug data exchange channel. Scan chain 3 reads and writes the register file of the processor core. Scan chain 4 accesses uniform peripheral address spaces through the clock domain converter. The uniform address spaces cover the external memory and the peripheral registers. The applications developers are able to easily acquire the data they want utilizing four debug scan chains.

The debug unit compares the instructions and data accessed by the processor core with the content of breakpoint registers preset by the developers. While they match, the debug unit outputs the breakpoint match signals, and then the system enters the debug state. There exist three sets of breakpoint registers in debug unit. Thus the applications developers can simultaneously set three different breakpoints or watchpoints, and the chip will come to a halt in the proper place of the program so long as at least one of the three breakpoints or watchpoints is encountered.

The debug data exchange channel performs the on-line communication between the processor core and the external debugger. Because the processor and the scan chains respectively belong to two different clock domains, two FIFO (First In First Out) queues are introduced to solve this problem. It provides a good mechanism for the debuggers to monitor the

real-time status of programs without disturbing the programs' normal execution.

The debug controller tightly couples with the processor core. It receives the match signal of breakpoints and watchpoints sent by the debug unit, and determines the state of the system. The main function of this module is to ensure the right results of programs' execution. It consists of the pipeline blocking logic, the processor's status preserving and resuming logic, the pipeline flushing logic, the cache writing back and invalid mechanism, a breakpoint instruction, etc. Fig. 2 shows the pipeline blocking logic. The pipeline runs in order while the "CanGo" signals of all stages are valid, otherwise the pipeline stalls. The BKPT_Go or WHPT_Go signal will be invalid if at least one of the breakpoints or watchpoints is encountered. Then the pipeline is blocked under the control of the cascade AND gates.

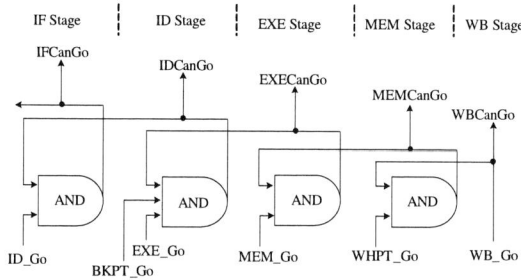

Fig. 2. Pipeline Blocking Logic

The clock switch unit changes the system clock when the chip enters or exits the debug state. The system clock is converted to test clock while the chip enters the debug state and processor clock while the chip exists the debug state. Although the gap of frequency between test clock and the processor clock is very wide and the two clocks are asynchronous, this module can well avoid the clock glitch, which is crucial for the whole chip. It is carefully designed and ensures the timing closure.

III. DEBUG SUPPORT FOR SCALABILITY

Since scalability and reusability are the essential features of System-on-Chip, the debug architecture should meet the demand of the features. In this section, we describe the debug components which support the scalability and reusability of SoC. These components consist of the corresponding debug scan chains and a module called a clock domain converter.

A. Debug Scan Chains

The software developers want to be aware of the status and results of instructions' execution. Therefore, it is necessary that the debugger can not only watch the register file and external memory space, but also modify the content of the whole memory. Then, we design two scan chains including the scan chain 3 and scan chain 4 to observe the register file and the memory space. These debug chains can strongly enhance the operability and controllability of the programs,

1550-4093/07 $25.00 © 2007 IEEE

and help the software developers to locate the faults rapidly and accurately.

However, scalability and reusability are the main features of System-on-Chip. When some peripheral IP cores are added or removed, the debug architecture should not change a lot. In general, the peripherals are connected to the on-chip bus in a slave mode. Therefore, we design a scan chain as a master module connected to the bus. This scan chain can send the address, data and control signals to the slave peripherals, and access the content of peripheral address space through the master port. Fig. 3 shows the structure of scan chain 4. There are three fields in scan chain 4, including the data, address, and control signals, which correspond to the signals of on-chip bus. To solve the problem of the increasing address bits while new IP cores are integrated, the address filed contains a few redundant bits. This simple mechanism meets the requirement of the scalable feature for SoC.

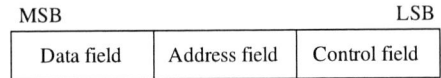

Fig. 3.   Structure of Scan Chain 4

B. Clock Domain Converter

The function of IP cores differs in thousands ways. We do not know which IP will be integrated into SoC. Some IP cores must work on a certain frequency. In other words, they cannot be under the control of test clock, because the frequency of test clock is very low. For example, since the sdram must refresh at intervals, the sdram memory controller should work on a fixed frequency, and cannot switch to test clock. There not exists the synchronous relation between the sdram clock and the test clock. Thus a module called a clock domain converter is designed to address this problem. This module entails the scalability solution more universal.

We implement the clock domain converter in a typical SoC chip-EStar. The on chip bus of EStar is similar to Wishbone [6] and AMBA [7]. These buses work in the mode of handshake. Fig. 4 shows the timing relation of main control signals. It is a procedure of the scan chain 4 writing data to the peripheral address space. To clearly describe the timing relation, some signals, such as address, write enable signals and so on, are omitted in this figure.

In Fig. 4, four signals above are the clock, data, request and acknowledge signals of scan chain 4. The four signals below are the clock, strobe, acknowledge, data signals of the on-chip bus. The last signal called ocb_ack is used to synchronize two acknowledge signals belonging to two different clock domains. It ensures that the scan chain controlled by the test clock can sample the acknowledge signal of the on-chip bus. When the scan chain 4 needs to write data to one of the peripherals, it prepares the output data and sends the request signal. While the strobe signal of on-chip bus samples the request signal, the peripheral reads the data and sends the acknowledge signal.

Then the ocb_ack signal is valid. The request and data signals of the scan chain 4 will not invalid until the ocb_ack signal is sampled by sc_ack_o signal. In this way, this module succeeds in communicating between the scan chain and the peripheral IP cores belonging to two asynchronous clock domains.

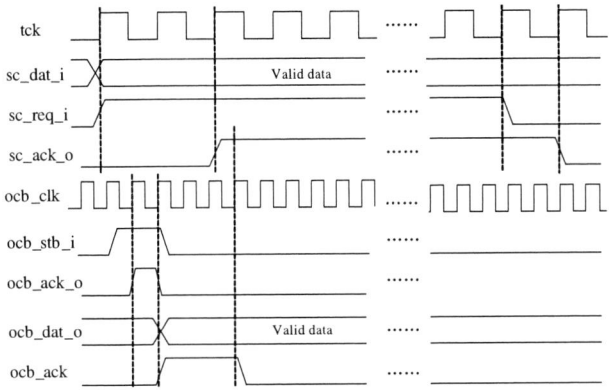

Fig. 4.   Timing Relations of Signals

IV. ON-LINE COMMUNICATION

It is necessary and convenient for application developers to understand the status of the processor real time without interrupting the programs' execution and switching to the debug state. Consequently, on-line communication is an indispensable function of the debug architecture. We design a module called a debug data exchange channel to satisfy the requirement. It provides a good framework to communicate between the processor and the debugger without stopping the program and entering debug state. The gap of frequency between the test clock and processor core clock is very wide, as much as order of magnitudes. Furthermore, the two clocks are asynchronous. Therefore, the data cannot be transmitted straight. The FIFO structure is introduced in the debug data exchange channel to solve this problem.

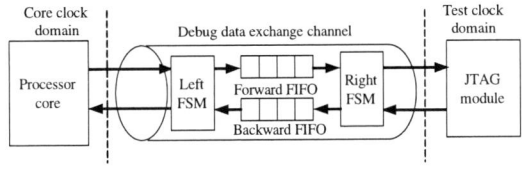

Fig. 5.   Structure of Debug Data Exchange Channel

We implement two FIFO queues in the debug data exchange channel. Fig. 5 is the structure of debug data exchange channel. Every FIFO transmits data in single direction. The depth of each FIFO queue is four. One queue is called a forward FIFO, through which the processor sends data and the debugger receives data. The other queue is called a backward FIFO, through which the debugger sends data and the processor receives data. The reading and writing operation of two FIFO queues are respectively controlled by left FSM and right FSM.

A status pointer called "FULL" indicates the queue full and data cannot be written any more. Another status pointer called "EMPTY" denotes the FIFO empty and no data can be read. In conclusion, two FIFO queues effectively avoid the influence of different clock domains. The debug data exchange channel transmits data accurately and efficiently between the processor and the debugger.

V. IMPLEMENTATION

To evaluate the performance of the debug architecture, it is implemented in a typical SoC chip called EStar. The process technology of EStar is 0.18 μm CMOS. Table I provides the resource utilization of each module. In Table I, the second column is the number of gates, and the area of each module is listed in the third column.

TABLE I

RESOURCE OF DEBUG MODULES

Module	Gates	Area (μm^2)
Debug unit	2540	90961.921875
Debug data exchange channel	507	16165.125485
Clock domain converter	213	11529.292969
Clock switch unit	42	2373.382721
Debug controller	826	25832.457836
JTAG controller	245	4859.893789
Scan chain 1	201	8585.439453
Scan chain 2	296	14034.053711
Scan chain 3	800	40585.312500
Scan chain 4	487	22263.568359

Table II gives the performance analysis of every unit. In Table II, the second column is latency of the critical path. The operation condition of static timing analysis is "WORST" condition, that is to say, the core supply voltage is 1.62 volt, and the I/O supply voltage is 3.0 volt, and the junction temperature is $125^{o}C$, and the process derating factor is 1.2.

The last two columns in Table II present power dissipation of each module. The constraints of power estimation are that the switch probability of logic signals is 20%, and the switch probability of clock signals is 200%, and the frequency of processor core clock is 200MHz, and the frequency of peripheral clock is 50MHz, and the frequency of test clock is 20MHz in the third column, and the frequency of test clock is 40MHz in the fourth column. The frequency of peripheral clock is one fourth of the processor core clock, and they are synchronous, that is to say, the peripheral clock is derived from the processor core clock by a frequency divider. Since the relation of many clocks in the clock switch unit is complex, it is difficult to exactly evaluate the values of timing and power. Thus in this table the timing of the clock switch unit is left blank, and the power is given an estimated value. Moreover, this table shows the power dissipation of the debug unit and the debug controller while they work in the debug state. When

they work in normal state driven by the processor core clock at 200MHz, the power dissipation of the debug unit and the debug controller respectively are 39.881mW and 24.086mW.

TABLE II

PERFORMANCE OF DEBUG MODULES

Module	Timing (ns)	Power (mW)	
Debug unit	2.16	4.002	7.009
Debug data exchange channel	1.18/1.10[1]	2.501	2.948
Clock domain converter	1.12/1.04[2]	0.996	1.270
Clock switch unit		<1	
Debug controller	1.98	2.435	4.935
JTAG controller	1.86	0.698	1.416
Scan chain 1	0.67	0.497	1.002
Scan chain 2	0.83	0.830	1.658
Scan chain 3	0.83	2.427	4.878
Scan chain 4	0.83	1.324	2.642

[1]The former value is critical path latency of the part driven by core clock, the latter is latency of the other part driven by test clock.
[2]The former value is critical path latency of the part driven by peripheral clock, the latter is latency of the other part driven by test clock.

From Table I and II, we arrive at the conclusion that each debug module occupies a few gates and small area. The percentage of the total area of debug modules is less than 2.5% of the chip area. The latency of critical path is small. Every unit achieves timing closure. Thus the debug architecture is well reusable. The power dissipation of each module is very small. Furthermore, when the chip is not in debug status, there is only static power dissipation for most debug logic, which even can be omitted. In a word, the debug architecture has high performance at the cost of few resources and area.

TABLE III

PERFORMANCE OF DEBUG OPERATIONS

Debug Operation	Cycles	Time (μs)	
Set three breakpoints	500	25.0	12.50
Set three watchpoints	500	25.0	12.50
Set breakpoint instruction	240	12.0	6.00
Start step operation	160	8.0	4.00
Read total register file	8640	432.0	216.00
Write total register file	8640	432.0	216.00
Read a word of memory	220	11.0	5.50
Write a word of memory	220	11.0	5.50
Break operation	30	1.5	0.75

Table III gives the performance of all basic debug operations. The second column is the number of cycles for each operation. The size of processor register file is 1K bits (32*32bits). The third column shows the total time of every operation when test clock is work at 20MHz frequency. The fourth column is the whole time of each operation driven by the test clock at 40MHz. The time consumption is at the level of microsecond, that is to say, it is even imperceptible for the the developers of software applications.

VI. CONCLUSION

Detecting and locating the faults in the software applications is difficult, tedious and time-consuming. On-chip debug is an important technique to solve the problem. Reusability and scalability are the essential features of SoC. Therefore, on-chip debug must adapt to the demand of reusability and scalability. In this paper, we present the novel debug architecture for scalable SoC. We design several special scan chains and a clock domain converter to meet the scalable requirement. On-line communication is very important for developers to locate errors in the programs. Thus we design a module called a debug data exchange channel to communicate between external debugger and the processor. The debug architecture is implemented in a typical single-core SoC chip. The results of performance analysis indicate that the debug architecture not only is well reusable but also has high performance at the cost of few resources and area. The future work is to further add the debug functions and develop the debugger to satisfy the better utility.

ACKNOWLEDGMENT

The authors would like to thank Shujing Gao for discussion and advice which helped to implement the debug architecture. This work is partially supported by the National Science Foundation of China (NSFC) under grant No. 60603088.

REFERENCES

[1] A. Mayer, H. Siebert, and K.D. McDonald-Maier, Debug Support, alibration and Emulation for Multiple Processor and Powertrain Control SoCs, Proceedings of the Design, Automation and Test in Europe Conference and Exhibition (DATE'05), March 2005.

[2] Andrew B.T. Hopkins, Klaus D. McDonald-Maier, Debug Support Strategy for Systems-on-Chips with Multiple Processor Cores, IEEE Transactions on Computers, vol. 55, no. 2, pp. 174-184, Feb. 2006.

[3] E. Moerman, S. Bocq, J. Verfaillie, Debug architecture for System on Chip taking full advantage of the Test Access Port, the Eighth IEEE European Test Workshop (ETW'03), May 2003.

[4] A. Mayer, H. Siebert, A. Kolof, and S. el Baradie, Debug Support for Complex System-on-Chips, CMP media LLC, Embedded Systems Conference, April 2003.

[5] Institute of Electrical and Electronics Engineers, IEEE Standard Test Access Port and Boundary-Scan Architecture Document, Number: IEEE 1149.1-2001, Jun. 2001.

[6] Silicore Corporation, WISHBONE System-on-Chip Interconnection Architecture for Portable IP Cores Revision: B.2, http://www.silicore.com, Oct. 2001.

[7] ARM Corporation, AMBA Specification Rev 2.0, http://www.arm.com/, May 1999

Abstraction and Refinement Techniques in Automated Design Debugging

Sean Safarpour, Andreas Veneris
Department of Electrical and Computer Engineering
University of Toronto, Toronto, Canada
{sean, veneris}@eecg.toronto.edu

Abstract— Verification is a major bottleneck in the VLSI design flow with the tasks of error detection, error localization, and error correction consuming up to 70% of the overall design effort. This work proposes a departure from conventional debugging techniques by introducing abstraction and refinement during error localization. Under this new framework, existing debugging techniques can handle large designs with long counter-examples yet remain run time and memory efficient. Experiments on benchmark and industrial designs confirm the effectiveness of the proposed framework and encourage further development of abstraction and refinement methodologies for existing debugging techniques.

I. INTRODUCTION

Functional verification of today's VLSI designs is a critical and time consuming task. The processes of verifying the functional correctness of a design, determining the source of the potential error(s) and correcting those errors, can take up 70% of the overall design time [1], [2]. While there exists a plethora of methodologies for verification (i.e. error detection), there is fewer work dedicated towards debugging, that is error localization and correction [3].

Today, the verification and design engineers have the daunting and tedious task of analyzing the design, its specifications and the incorrect response from the simulation traces (counter-examples) to determine the source of errors. For real-life industrial designs such as microprocessors and DSP components, experience shows that traces can often be over tens of thousands of clock cycles long, a fact that makes debugging even more challenging [1], [4]. It is reported that finding the source of error(s) can take up to 50% of the overall verification task, a considerable contributor to the verification bottleneck [1]. Therefore, cost-effective automated debugging methodologies are of great importance to the academic and industrial communities.

Currently, automated design debugging approaches are based on simulation, symbolic, or constraint satisfaction techniques [5], [6], [7]. Most of these approaches use information from the erroneous design, the input logic values, and the expected output logic values to return a set of suspect gates [7]. In sequential design debugging, the circuit representation is often replicated or unrolled for all clock cycles to model the error through time [6]. Clearly, this debugging practice can result in excessive memory and run-time requirements even for modest size designs with hundreds of clock cycle traces.

Reducing the memory demands for sequential debuggers is critical in making existing debugging methodologies practically viable. Current memory reduction techniques partition the problem into subproblems that are solved sequentially [7], they trade time for space by formulating the problem as a Quantified Boolean Formula satisfiability instance [8], and they reduce the length of the traces using formal techniques [9], [10]. Although these methods can be effective in decreasing memory requirements, the problem of debugging large industrial designs remains intractable.

This work introduces a new methodology for design debugging based on the formal concepts of design abstraction and refinement [11], [12]. It does not propose a new debugging method but it presents a novel framework that existing debugging techniques can utilize. Under this new framework, an abstract model of the design is first created to undergo debugging. Since this representation contains less logic than the original one, the size of the problem may be reduced considerably in favor of debugging. The benefits are reduced memory requirements and potentially shorter run times when compared to debugging the original design. Since the abstract model contains fewer state elements than the original one, it may lead to shorter traces when state matching trace reduction techniques are used [9]. Debugging an abstract model can sometimes return abstracted state elements as error sources. In these cases, a refinement procedure is proposed that replaces some of the abstracted variables with the original state elements. Furthermore, the proposed debugging methodology guarantees correctness, where the solutions found under the framework are also solutions in the concrete design. Finally, the completeness of the framework is ensured as no error sources remain undetected for a given set of test vectors.

Given the impact of abstraction and refinement techniques in model checking, it is natural to expect similar results in design debugging. Indeed, experiments confirm the practical benefits of the proposed framework as considerable memory reductions of over 60% and speed-ups of over 4.5X are observed on a set of benchmark and industrial designs while preserving the resolution. These results encourage further research in abstraction and refinement methodologies as an aid to existing debugging techniques.

This paper is organized as follows. Section II provides background information relating to debugging as well as abstraction and refinement. The proposed abstraction and refinement debugging framework is presented in the Section III. The empirical results are provided in Section IV while Section V concludes this work.

II. PRELIMINARIES

This section presents terminology used in the paper. It provides a brief background on design debugging using Boolean satisfiability (SAT) as well as an introduction to abstraction and refinement techniques in model checking. Although SAT-based debugging is used to explain various theoretical concepts in this paper and is used as the debugging engine in the experiments, the proposed framework is not confined to any unique aspects of this debugging technique.

A. Background

Given a set of vectors V for which a circuit (or netlist) C demonstrates an incorrect behavior, the objective of design debugging is to find the gates that may be responsible for this incorrect behavior [5]. In the context of this paper, the set of vectors V include the initial state value, the sequence of primary input values and the *correct* or *expected* primary output values for every clock cycle or time frame. In other words, the specifications for the erroneous circuit act as a "black box" without knowledge about its internal structure. The set V can be derived from a simulation trace or from a formal verification engine. The terms vector, trace, and counter-example are used interchangeably in this paper.

A circuit C is composed of a set of primary inputs $x_1, x_2, ...$, primary outputs $y_1, y_2, ...$, primitive gates $l_1, l_2, ...$, and state elements $q_1, q_2, ...$ such as flip-flops or latches. An interconnect is referred to by the name of its driving gate. For example, the wire connecting the output of gate l_i to an input of gate l_j is simply referred to as l_i.

B. Debugging with Constraint Satisfaction

The constraint satisfaction problem in SAT-based debugging is generated by adding extra logic to the erroneous circuit C, converting the new circuit into Conjunctive Normal Form (CNF), replicating and constraining the CNF for every vector and time frame in V [7]. The constraint or CNF problem is solved by a SAT solver which returns a set of gate locations where a *correction* (function change) can produce the expected outcome captured in V.

To model the corrections in C, a multiplexer m_i is added for every gate (and primary input) l_i. The output of this multiplexer, m_i, is connected to the fanouts of l_i while l_i is disconnected from its fanouts. This construction has the following effect: when the select line s_i of a multiplexer is inactive ($s_i = 0$), the original gate l_i is connected to m_i, otherwise, when $s_i = 1$ a new unconstrained primary input w_i is connected. Figure 1 illustrates the above transformation for a combinational circuit.

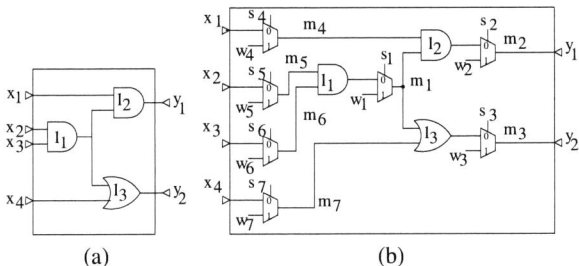

Fig. 1. Extra logic for SAT-based debugging

A potential correction on line l_i is indicated when the select line s_i is assigned to 1 under which condition the correction value is stored in w_i. The SAT solver can assign any value $\{0,1\}$ to the s_i and w_i variables such that the CNF satisfies the constraints applied by the vectors V. To force the SAT solver to find a specific number N of error locations, further logic is added to activate at most N select lines. Thus for $N = 1$, a single s_i is set to 1 which *corresponds* to candidate error location l_i. In the following, an erroneous gate location is referred to as an error source and an error stemming from N distinct locations is called an N-tuple error.

It is known that many *equivalent* errors may exist for a set of vectors and for a fixed design error model [6]. Intuitively, this is true because there may be more than one way to synthesize and correct a design. In this paper, we say that the debugging procedure is not *complete* unless all equivalent error sources are found. This is performed iteratively by finding a solution to the CNF problem, adding it subsequently to the CNF as a blocking clause and solving the problem again until no more solutions are found.

C. Abstraction and Refinement

Abstraction and refinement techniques are used readily in model checking to mitigate the exponential nature of the underlying state space [11], [12], [13]. Roughly speaking, an *abstract model* is derived by removing some state elements from the original or *concrete* design using some abstraction function $\hbar$. The reduced number of state elements result in fewer states to consider when verifying properties.

An abstract model can be derived by the following simple steps:

1) Use the abstraction function $\hbar$ to remove some state elements from the concrete design.

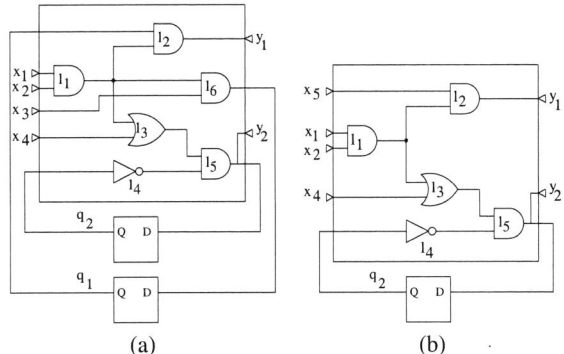

Fig. 2. Abstracting q_1 from a circuit

2) Remove all the combinational logic that are only in the transitive fanin of the abstracted state elements.

3) For each removed state element, introduce a primary input and connect it to the fanout of a removed state element.

For example, Figure 2 (a) and (b) illustrate a circuit before and after abstracting the state element q_1, respectively.

For safety properties, if model checking determines that a property holds in the abstract model, then it must also hold in the concrete design [11]. However, if a property does not hold in the abstract model, then the corresponding counter-example may or may not hold in the concrete design. If the counter-example is not valid on the concrete design it is said to be *spurious* [11]. In this case, the abstract model is *refined* by reverting some of the abstracted state elements and continuing the model checking process.

III. DEBUGGING WITH ABSTRACTION AND REFINEMENT

This section proposes a new framework for debugging using abstraction and refinement. The initial formulation is presented in Section III-A. In Section III-B, the occurrence of unjustifiable solutions is elaborated and the methodology is re-formulated to prevent them. In Section III-C, this framework is extended for completeness.

A. Basic Construction

The abstraction and refinement methodology first creates an abstract model. This model and its corresponding set of vectors V are used to generate a new debugging problem instance. If debugging is formulated as a constraint problem described in Section II-B, the problem is solved by a SAT solver. Next, the solutions returned by the SAT solver must be verified on the concrete design to determine whether they are unjustifiable or spurious. Spurious solutions are used to refine the abstract model and the process is repeated with the new model. In the following, we describe this process in detail.

An abstract model C' is constructed by removing (i.e. abstracting) the state elements and replacing them with primary inputs. Which and how many state elements are selected for removal is determined by the abstraction function $\hbar$. Once the state elements are abstracted, all logic in their transitive fanin is also removed.

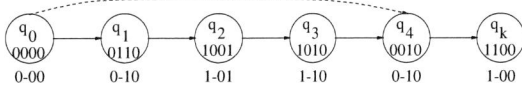

Fig. 3. Reduced trace V' due to abstraction

Since the abstract model may have fewer state elements than the concrete model, a set of more compact traces V' can be obtained through state matching [9], [10]. State matching procedures can remove redundant state transitions between repeated states or they can employ formal techniques to determine the minimum trace length achievable.

1550-4093/07 $25.00 © 2007 IEEE

Algorithm 1 Basic Abstraction and Refinement Debugging

```
 1: Solutions = ∅
 2: C' = abstract(C)
 3: while (1) do
 4:    V' = trace_reduction(C', V)
 5:    New_sols = debug(C',V')
 6:    if ( New_sols = ∅) then
 7:       return Solutions
 8:    end if
 9:    Solutions = Solutions ∪ New_sols
10:    for all  Sol ∈ New_sols do
11:       if ( unjustified_solutions(Sol, C, V')) then
12:          Solutions = Solutions \ Sol
13:       else if ( spurious_solutions(Sol, C')) then
14:          C' = refine(Sol,C')
15:          Solutions = Solutions \ Sol
16:       end if
17:    end for
18: end while
```

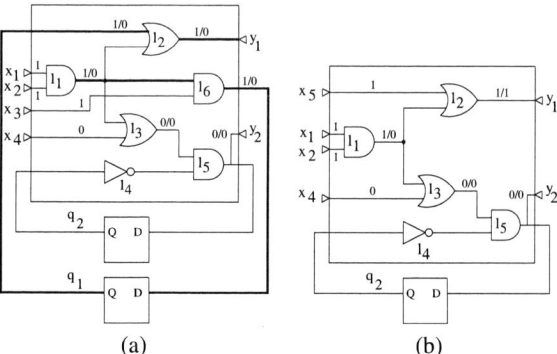

Fig. 4. Effect of unconstrained variables

As an example consider Figure 3 where a state transition diagram is used to illustrate an error trace from state q_0 to q_k. In the original trace, no trace reductions are possible through state matching. However, after the second state element is removed (through abstraction) the states q_1 and q_4 can no longer be differentiated. The state values after abstraction are shown under each node in Figure 3. As a result, a "short-cut" can be taken in the trace from state q_0 to state q_4 as illustrated by the dashed line. Similar to many trace reduction techniques, the compacted traces V' must be tested to determine whether the error(s) are still observable [10].

Next, the abstract model and the compacted traces V' are used to formulate the debugging problem. Since the abstracted model contains less logic and has potentially shorter traces than the concrete design, the size of the CNF is expected to be smaller in terms of clauses and variables and thus require less memory. As described in Section II-B, the SAT solver returns a set of location tuples which are potential error sources. These potential error sources represent locations where a correction (*i.e.,* some function change) can be applied which results in a correct behavior of the abstracted model as dictated by the reduced set of vectors V'. The SAT solver also assigns logic values to all abstracted variables which are now unconstrained primary inputs in the abstract model.

Although viable for the abstract model, the solutions returned by the SAT solver may not be viable for the concrete design. More specifically, the logic value assignments made to the abstract variables by the solver may not be justifiable in the concrete design where these variables are constrained by their original fanin logic.

Definition 1 *Debugging an abstract model results in an* unjustifiable *solution if the logic value assignments cannot be justified for the corresponding variables in the concrete design.*

To verify whether the solutions are justifiable, a constraint problem is formulated using the concrete design, C, the solutions returned by the debugger, and the reduced traces V'. This problem is provided to a SAT solver where an unsatisfiable result determines that the abstracted variable logic assignments are unjustifiable according to Definition 1.

Definition 2 *Debugging an abstract model results in a* spurious *solution if any of the error tuples returned correspond to an abstracted variable.*

According to Definition 2, spurious solutions do not provide enough information about the error sources since the abstracted variables have their fanin logic removed from the original design. In this case, the model is *refined* using these abstracted variables so that the non-abstracted error sources can be found. The debugging continues with this newly refined model until all equivalent error sources are found.

Algorithm 1 illustrates the basic abstraction and refinement procedures described thus far. On lines 2 and 4, the initial abstract model C' and the reduced set of vectors V' are generated. Line 5 calls the debugger to search for solution tuples while the unjustifiable solutions are removed on line 12. For the remaining solutions, spurious ones are filtered out and the abstract model is refined as shown on lines 14-15. This process is repeated until no new solutions are found.

As mentioned earlier, algorithm 1 is not restricted only to SAT-based debugging. For instance, Binary Decision Diagram (BDD) based debugging methods [6] can be employed by building a BDD representation of the abstract circuit. Similarly, simulation-based techniques [6], can perform simulation and path-tracing on the abstract model as they search for solutions.

B. Guaranteeing Correctness

In the formulation of Section III-A, by leaving the abstract variables unconstrained in the problem CNF, the SAT solver may trivially assign them logic values so that the erroneous abstract model produces the correct response. In fact, the SAT-based debugging formulation of Section II-B may be satisfied without activating any of the multiplexer select lines (i.e. $\forall i, s_i = 0$) or when $N = 0$. The following example illustrates this situation.

Example 1 *Figure 4 (a) shows a concrete design with an error on gate l_1 which forces the output to 0. The correct/erroneous value of 1/0, shown in bold, propagates from gate l_1 through the flip-flop q_1 and to the primary output y_1. Notice that the primary input values remain constant for both time frames. When the state element q_1 is abstracted and left unconstrained, the SAT solver can assign this new input x_5 to 1 which will produce the correct/erroneous outcome 1/1 as shown in Figure 4 (b).*

The above example shows that the SAT solver can satisfy the problem without activating any multiplexer select lines. As a result, when $N \geq 1$, some solutions returned by the SAT solver are *unjustifiable* as stated by Theorem 1 below.

Theorem 1 *There exist automated debugging problem instances with unconstrained abstracted variables such that the solutions to the problems are unjustifiable.*

Proof: As shown by Example 1, leaving abstracted values unconstrained can result in satisfying the constraint problem with $N = 0$. Since the debugger finds locations that can correct the abstract design with $N \geq 1$, all N-tuple locations in the design qualify by setting any s_i variable to 1 and assigning the value of the l_i variables to the w_i variables. In effect the solutions simulate the behavior of the abstract circuit which is error-free under the current variable assignment. Since there is an error in the erroneous circuit by definition, the solutions for the abstract model must be unjustifiable in the concrete design. ∎

1550-4093/07 $25.00 © 2007 IEEE

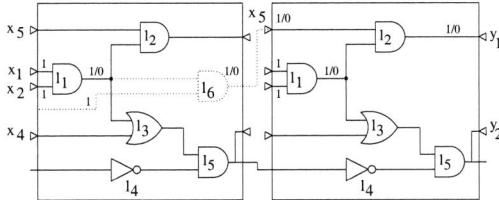

Fig. 5. Abstract model for two time frames

The consequence of Theorem 1 is that the process of determining whether solutions are unjustifiable, presented in Algorithm 1, may not be practical since all N-tuple solutions must be verified to work on the concrete design. Theorem 2 states that unjustifiable solutions can be prevented altogether.

Theorem 2 *Constraining abstracted variables to the values of the corresponding concrete state elements from sequential simulation using the initial state and input values in V' prevents unjustifiable solutions from occurring.*

Proof: The objective of this proof is to show that the abstract model can be restricted sequentially to behave like the concrete model and thus prevent unjustifiable solutions from occurring. There exist a well defined sequence of state transitions from the initial state to the state where the erroneous behavior is witnessed that can be observed through the simulation of C with the input and initial state vectors V'. By applying the logic values of the concrete state elements onto the corresponding abstracted variables, the latter will be constrained to the values in V'. As a result, both the concrete design and the abstract model will contain the same constraints except that one is enforced by logic circuitry while the other is enforced by logic values. Since unjustifiable solutions do not occur in the concrete design by definition, they will not occur in the abstract model either. ∎

By Theorem 2, unjustifiable solutions are prevented by constraining the abstracted variables to the values of their concrete counterparts. Thus *correctness* is guaranteed since all solutions for the abstract model are also solutions for the concrete design. Next, the proposed methodology is extended to guarantee that all equivalent error sources are found.

C. Guaranteeing Completeness

SAT-based debuggers such as those described in Section II-B can find all actual and equivalent errors for a given value of N. In the methodology described in Section III-B, it is not the case that a single gate-level error is found at $N = 1$. In other words, a set of m errors in the concrete design may be mapped onto a set of n errors in the abstract model, where $n > m$, as the following example illustrates.

Example 2 *Consider the abstract circuit in Figure 2 (b) unfolded over two time frames as illustrated in Figure 5. For clarity, the abstracted logic l_6 is shown in dashed lines. Notice that the error from gate l_1 does not directly propagate to output y_1 but its effect is captured in the abstract variable x_5. For $N = 1$ the SAT solver returns the single equivalent error location l_2. Assuming that the design is analyzed and it is concluded that l_2 is not the error source, the real source of error goes undetected. However, if N is incremented to 2, then the pair l_1 and x_5 is found as a solution. By refining the abstract variable x_5 to q_1 and solving the debugging problem again with $N = 1$, the single error location l_1 is found.*

Theorem 3 states that the process outlined in Example 2 finds all equivalent error locations and is thus complete.

Theorem 3 *The debugging procedure that performs the following steps is complete for some value of $maxN$.*

1) *Perform debugging for N-tuple errors using the abstract model*
2) *If an abstracted variable is returned as an error location, refine the model and set $N = 0$*
3) *Increment N by 1*
4) *Go to (1) unless $N > maxN$*

Algorithm 2 Complete Abstraction and Refinement Debugging

```
 1: Solutions = ∅, N = 1
 2: C' = abstract(C)
 3: while (1) do
 4:     V' = trace_reduction(C', V)
 5:     Const = extract_constraint(C', V')
 6:     New_sols = constrain_and_debug(C',V',N,Const)
 7:     Solutions = Solutions ∪ New_sol
 8:     for all Sol ∈ New_sols do
 9:         if (spurious_solutions(Sol, C')) then
10:             C' = refine(Sol, C')
11:             Solutions = Solutions \ Sol
12:             N = 0
13:         end if
14:     end for
15:     N = N + 1;
16:     if (N > maxN) then
17:         return Solutions
18:     end if
19: end while
```

Proof: Since at some point N will equal the number of errors mapped in the abstract design n, all the equivalent errors that map into n-tuples or fewer error sources will be found. If any of these locations correspond to abstract variables, then the abstract model is refined and those variables are replaced with their corresponding concrete state elements. The new abstract model is then provided to the debugger which starts the search with $N = 1$. Since some previously abstracted variables no longer exist in the new abstract model, previous solutions at $N = n$ will be found at $N \leq n$. This process continues until no new solutions are found which guarantees that all error sources are found for $maxN = n$, an event that guarantees completeness. ∎

Empirical results from Section IV show that for single errors, $maxN = 3$ is large enough to find all equivalent error locations. Furthermore, not much time is spent on debugging problems for which there exists no solutions for a particular N. Algorithm 2 illustrates the overall abstraction and refinement based debugging methodology that guarantees correctness and completeness. It is similar to Algorithm 1 with the following differences. On lines 5 and 6 simulation values are extracted and used to constrain the abstract variables. Spurious solutions are refined and N is reset on lines 10-12 while N is incremented and checked to be less than $maxN$ on lines 15-17.

IV. EXPERIMENTS

This section presents the experiments conducted to evaluate the effectiveness of the proposed framework. The debugging problems are generated using a sample of four ISCAS'89 circuits, four ITC'99 circuits and four industrial circuits from OpenCores.org [14]. The erroneous circuits are created by changing the type of a single gate at random. An average of 10 traces are obtained per problem through pseudo-random simulation of the correct and buggy circuits until a different outcome is observed. The automated debugger used is a sequential SAT-based debugger similar to [7]. The experiments are conducted on a 2.66GHz Intel Xeon processor with 2 GB of memory with a timeout of 7200 seconds for each SAT problem.

Before starting the debugging process, a simple trace compaction procedure is always performed. This procedure first builds a graph of the visited states, it then connects edges between repeated states and applies Dijkstra's shortest path algorithm from the initial state to the final state [15]. More effective trace compaction algorithms can be applied for better results [9], [10]. After the trace compaction, all traces are used as simulation stimulus to ensure that they still exhibit a different result for the correct and buggy designs.

In summary, Figure 6 shows the effects of abstraction on the logic size and trace lengths. Part (a) demonstrates that the logic size reductions appear to be linear with respect to the number of abstracted state elements. However, in part (b), experiments show that significant trace length reductions are not observed until a certain threshold is

---

1550-4093/07 $25.00 © 2007 IEEE

TABLE I

PROBLEM INFORMATION AND STATISTICS FOR STAND-ALONE SAT-BASED DEBUGGING APPROACH

circuits	# gates	# FF	# clk	# red. clk	# cls (K)	mem (MB)	time/err (s)	# err	total (s)
b04	711	66	516	335	2422	1132	740.0	9	6660.0
b08	200	21	21	20	274	82	3.8	4	15.2
b12	1140	121	40	19	1492	449	165.9	5	829.5
b14	6028	245	54	54	memout	> 2000	-	-	-
s1488	693	6	104	5	214	42	1.6	9	14.4
s5378	3222	179	3	3	554	105	13.1	3	39.3
s13207	9442	669	2	2	1415	227	70.1	9	630.9
s35932	21147	1728	75	8	3563	696	431.1	16	6897.6
div_su	1528	126	9	6	607	109	12.4	64	793.6
rsdecoder	10629	521	2	2	2043	301	120.1	9	1080.9
spi	2027	90	20	18	2763	582	391.3	3	1173.9
ac97	15166	1452	30	30	memout	> 2000	-	-	-

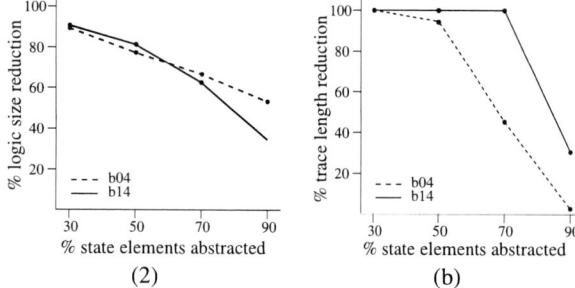

Fig. 6.   Logic and trace reduction vs. flip-flops abstracted

reached. This threshold appears to be over 50% for b04 and over 70% for b14. Thus for large problems where memory is a major concern, a more aggressive approach, where over 70% of state elements are abstracted, may be desirable.

Table I presents a summary of the debugging problems as well as some insightful performance statistics for debugging the concrete circuits using the conventional SAT-based debugger. Later, these results are contrasted with those of the proposed debugging framework. Columns 1, 2 and 3 present the circuit name, number of gates, and number of flip-flops (state elements) in each circuit. Columns *clk* and *red. clk* show the average length of the traces before and after the trace compaction, respectively.

The next five columns summarize the results of the debugger for each problem. In Columns *# cls* and *mem*, the number of clauses in thousands generated for each problem and the debugger's memory usage is presented. The number of equivalent errors found by the debugger for the given vectors as well as the average time required to find them are presented in columns *# err* and *time/err*, respectively. Finally, the total time required to find all the errors is presented in column *total*.

To cope with the size of the larger problems the CNFs are partitioned into bands and solved sequentially as described in [7]. For b14 and ac97 where the average reduced traces are 54 time frames and 30 time frames long, the problems still run out of memory. The proposed framework that uses abstraction is most beneficial for such memory intensive problems.

Table II presents the results of the proposed abstraction and refinement SAT-based debugging framework. For each problem, a random abstraction function is used such that between 40-50% of the state elements are abstracted, a conservative amount according to Figure 6. To allow easy comparison with Table I, the percentage of reduced logic, reduced of flip-flops, additional compacted traces, and overall reduced memory requirements are presented in columns 2-5, respectively. Looking across one row for problem b08, by abstracting 47.6% of the flip-flops, the logic is reduced by 26% and the trace length is reduced by an additional 65% which leads to an overall memory reduction of 60% versus the stand-alone debugger.

The largest problems in Table I, b14 and ac97, which ran out of memory previously are both solved under the proposed framework. On the average, the proposed methodology results in up to 60% memory reduction with average savings of 30% under a conservative abstraction approach.

The majority of problems in Table II do not benefit from additional trace compaction. This can be attributed to the fact that trace reduction is most effective for long traces since the probability of matching states is higher. In the experiments, the initial trace compaction process is able to reduce the traces considerably. For instance, the initial trace of circuit s1488 which is 104 clock cycles is reduced to only 5 clock cycles after compaction, thus further reductions are highly unlikely. For industrial traces of thousands of clock cycles derived from functional testbenches and not randomly, it is highly unlikely to reduce traces drastically by simple state matching techniques [9]. Therefore, trace reduction via abstraction may be more effective.

A summary of the run time results of the proposed framework is presented in columns 6-12 of Table II. In columns *time/err* and *# err* the average time required to find an error and the number of errors found are presented, respectively. It should be noted that when the number of errors are greater than those in Table I, it means that abstracted state variables are found as errors. In these experiments, if all equivalent error tuples are found (including the actual inserted error), then no refinement is performed. Note that in practice, not all equivalence errors are necessary since only the actual error is of interest to the designer. If the errors found by the proposed framework do not include all equivalent error locations (i.e. *# err* is smaller in Table II than Table I), then all spurious solutions must be refined.

In Table II, column *maxN* shows the maximum number of tuples searched until all equivalent errors are found. The debugging time for all searches prior to *maxN* is shown in the column *prev*. When refinements are necessary, the column *refine* presents the solve time for all subsequent refinement searches.

For many problems in Table II, the maximum error tuple found (*maxN*) is often greater than 1 but always less than or equal to 3. This signifies that a single error in the concrete design maps to 3 or fewer locations in the abstract model. The time required to determine that no solutions exist prior to *maxN* (*prev*) is always quite smaller than the average time required to find an error (*time/err*). Take b12 for instance, it takes on average 4.2 seconds to determine that no errors occur when $N < 3$ and 85 seconds to find each solution at $N = 3$. Relating these times to Algorithm 2, it means that the approach is quite effective since the majority of the time is spent in `constrain_and_debug` occurs when $N=maxN$ and not when $N < maxN$.

The total debugging time for the proposed approach is found by summing the product of *time/err* and *# error* with *prev* and *refine*. The resulting total run time is shown in column *total* and its improvement over Table I is shown in column *X impr*. When abstracting 40-50% of the state elements, not many refinement steps are necessary as most

1550-4093/07 $25.00 © 2007 IEEE

TABLE II
PERFORMANCE STATISTICS FOR ABSTRACTION AND REFINEMENT DEBUGGING FRAMEWORK

circuits	red. logic(%)	red. FF(%)	red. trace(%)	red. mem(%)	time/err (s)	# err	maxN	prev (s)	refine (s)	total (s)	X impr.
b04	20.5	45.4	0	**9.8**	530.0	12	3	11.0	0	6371	**1.04**
b08	26.0	47.6	65.0	**60.0**	0.2	12	3	0.1	0	3.35	**4.53**
b12	26.4	41.3	15.7	**24.9**	85.0	20	3	4.2	0	1704.2	**0.48**
b14	15.3	40.8	0	**> 46.0**	3740.2	2	2	42.0	0	7522.4	**-**
s1488	20.4	50.0	0	**11.9**	1.1	9	1	0	0	9.9	**1.45**
s5378	9.7	44.6	0	**37.1**	11.8	1	1	0	3.4	15.2	**2.58**
s13207	29.6	44.8	0	**31.7**	40.3	9	1	0	0	362.7	**1.73**
s35932	31.9	46.2	0	**34.9**	251.3	16	2	7.3	0	4028.1	**1.71**
div_su	34.0	39.6	0	**9.5**	5.9	32	3	2.2	396.8	587.8	**1.35**
rsdecoder	34.7	43.1	0	**22.9**	54.8	7	1	0	0	383.6	**2.81**
spi	37.6	44.4	22.2	**46.0**	101.2	1	1	0	303.6	404.9	**2.89**
ac97	41.2	48.2	0	**> 37.0**	365.6	2	1	0	0	731.2	**-**

TABLE III
SUMMARY OF B14 WHEN ABSTRACTING OVER 80% OF FLIP-FLOPS

step	red. logic(%)	red. FF(%)	red. trace(%)	mem(MB)	time/err(s)	err
Tbl II	15.4	40.8	0	1080	3740.0	2
abs	52.7	81.6	20.3	344	172.0	4
ref 1	50.2	80.8	20.3	378	225.1	3
ref 2	50.1	80.4	20.3	404	242.3	10

TABLE IV
SUMMARY OF AC97 WHEN ABSTRACTING OVER 96% OF FLIP-FLOPS

step	red. logic(%)	red. FF(%)	red. trace(%)	mem(MB)	time/err(s)	err
Tbl II	41.2	48.2	0	1260	1567.8	2
abs	89.7	96.4	33.3	555	365.6	2
ref 1	89.5	96.3	33.3	765	665.8	10
ref 2	89.4	96.2	33.3	773	664.0	6
ref 3	89.1	96.1	33.3	776	721.8	9

equivalent error locations are found in the abstract model. However, even for the cases where refinement is necessary, substantial run time improvement is observed. The only problem that demonstrates a performance decrease is b12 where four times more solutions are found in the abstract model versus the concrete design. Overall, performance improvements of up to 4.5X are observed with an average value of 2X across all problems. This increased efficiency can be attributed to the smaller size of the constraint problems which lead to easier CNFs for the SAT solver.

As observed in Figure 6 smaller problem sizes and shorter traces can be achieved with more aggressive abstraction than those of Table II. To demonstrate the effectiveness of the framework under a more aggressive abstraction strategy, the two largest problems b14 and ac97 are shown in Tables III and Tables IV with 80% and 96% of the state elements abstracted. For easy comparison, the first row of each table re-presents the problem properties of Table II. The following rows show the results after each abstraction and refinement steps until the specific injected error is found (not all equivalent errors as in Table I). For each table, column 1 describes whether the data is derived from Table II (*Tbl II*), from the initial abstraction (*abs*), or from a refinement step (*ref*). The remaining columns are labeled similarly to Table II.

As expected, when more state variables are abstracted, greater memory saving are attained and more refinement steps are necessary. However, along with the memory savings, more abstracted variables lead to much faster solve times per error. For instance, b14 requires 3740 second per error with 40% state abstraction while it requires only 172 seconds per error with 82% state abstraction.

It is interesting to notice the relatively small number of iterations necessary to find the injected error. More precisely, b14 and ac97 require only two and three refinement steps, respectively, before finding the errors. This small number of steps indicates that the appropriate variables are selected for refinement and that the debugger

is guided efficiently towards the errors after each step.

Overall, the proposed abstraction and refinement debugging framework demonstrates its effectiveness for large problems where conventional approaches may fail due to excessive memory and/or run-time requirements.

V. CONCLUSION

In this work, a novel debugging framework is proposed based on abstraction and refinement. The framework creates an abstract model which undergoes debug and demonstrates substantially reduced memory requirements. By formulating the problem carefully, the overall approach guarantees completeness and correctness. The experiments demonstrate that problems too large for a conventional approach are solved by the proposed framework. Furthermore, memory reductions of over 60% and run time improvements of over 4.5X are observed. Overall, the results encourage further work in the area of abstraction and refinement as an efficient platform for design debugging.

REFERENCES

[1] P. Rashinkar, P. Paterson, and L. Singh, *System-on-a-chip Verification: Methodology and Techniques*. Kluwer Academic Publisher, 1996.

[2] R. Drechsler, *Formal Verification of Circuits*. Kluwer Academic Publishers, 2000.

[3] Y. Yang, S. Sinha, A. Veneris, and R. Brayton, "Automating Logic Rectification by Approximate SPFDs," in *ASP Design Automation Conf.*, 2007.

[4] D. Appenzeller and A. Kuehlmann, "Formal verification of a PowerPC microprocessor," in *Int'l Conf. on Comp. Design*, 1995, pp. 79–84.

[5] M. Abramovici, M. Breuer, and A. Friedman, *Digital Systems Testing and Testable Design*. Computer Science Press, 1990.

[6] S. Huang and K. Cheng, *Formal Equivalence Checking and Design Debugging*. Kluwer Academic Publisher, 1998.

[7] A. Smith, A. Veneris, M. F. Ali, and A. Viglas, "Fault diagnosis and logic debugging using Boolean satisfiability," *IEEE Trans. on CAD*, vol. 24, no. 10, pp. 1606–1621, 2005.

[8] M. F. Ali, S. Safarpour, A. Veneris, M. Abadir, and R. Drechsler, "Post-verification debugging of hierarchical designs," in *Int'l Conf. on CAD*, 2005, pp. 871–876.

[9] Y. Chen and F. Chen, "Algorithms for compacting error traces," in *ASP Design Automation Conf.*, 2003, pp. 99–103.

[10] K. Chang, V. Bertacco, and I. Markov, "Simulation-based bug trace minimization with BMC-based refinement," in *Int'l Conf. on CAD*, 2005, pp. 1045–1051.

[11] E. Clarke, O. Grumberg, and D. Long, "Model checking and abstraction," in *Symposium on Principles of Programming Languages*, 1992, pp. 342–354.

[12] E. Clarke, A. Gupta, and O. Strichman, "SAT-based counterexample-guided abstraction refinement," *IEEE Trans. on CAD*, vol. 22, no. 7, pp. 1113–1123, 2004.

[13] P. Bjesse and J. Kukula, "Using counter example guided abstraction refinement to find complex bugs," in *Design, Automation and Test in Europe*, 2004, pp. 156–161.

[14] OpenCores.org, "http://www.opencores.org," 2006.

[15] T. Cormen, C. Leierson, and R. Rivest, *Introduction to Algorithms*. MIT Press, McGraw-Hill Book Company, 1990.

1550-4093/07 $25.00 © 2007 IEEE

Diagnosing Silicon Failures Based on Functional Test Patterns

Chia-Chih Yen[1], Ten Lin[1], Hermes Lin[1], Kai Yang[2], Tayung Liu[2], and Yu-Chin Hsu[2]

[1]Springsoft, Inc., Hsinchu, Taiwan

{chiachih_yen, ten_lin, hermes_lin}@springsoft.com

[2]Novas Software, Inc., San Jose, USA

{kyang, tyliu, ychsu}@novas.com

Abstract—**Identifying the root-cause of silicon failures is crucial for silicon debug and yield improvement. However, due to the low visibility of silicon data, root-cause identification tends to be a painful process. In this paper, we develop a systematic framework to diagnose silicon failures under functional test patterns. We propose a novel scan-dump approach to isolate critical cycles. Within the critical cycles, we apply logic-reasoning techniques including active-path-tracing (AP) and what-if (WI) analysis to automatically extract and rank failure candidates. We apply our framework on an industrial circuit and demonstrate the promising results.**

Index Terms—**Silicon debug, design for debug, fault diagnosis**

I. INTRODUCTION

As designs aggressively growing up and feature sizes steadily shrinking, it is not rare that the same functional test pattern passes in pre-silicon verification yet fails in post-silicon. Since the design errors or the imperfect manufacturing process can cause silicon failures, it is usually difficult to identify the effects. Furthermore, low visibility of silicon data prevents engineers understanding the internal behavior of silicon. To relieve the pain of silicon debug, a systematic method to localize the root cause of silicon failures is necessary and critical.

Typically, diagnosing silicon failures includes the following steps:

· Isolate critical cycles

· Rank failure candidates for further physical probing

· Confirm candidates using FIB (Focused Ion Beam)

The first step is to isolate the critical cycle. Critical cycle is the first cycle which the failure is activated and observed. By utilizing the Design-for-Debug (DFD) methodology [1]-[2], such as scan-dump, engineers can compare the observed responses and the pre-silicon simulation to determine which cycle firstly activates errors. The challenging is how to precisely scan-dump the given cycle under functional pattern testing. If the dumped data cannot guarantee to be the correct snapshot, the following diagnosis flow might produce wrong results.

The second step is to perform detailed failure candidate ranking within the isolated critical cycles. Many proposed diagnosis methods already address on this step [3]-[6]. However, most existing techniques are just applicable for combinational circuit. In practice, due to the capability of DFD, the isolated critical cycles may cross more than one internal clock cycle in functional test patterns. Therefore, sequential elements (flip-flop, latch…) and asynchronous signals (clock, control) must also be considered.

After root-causing the failure candidates, the third step is to use physical tools [7] such as FIB to directly confirm the failure candidates. Since utilizing physical tool is very time-consuming, the result of candidate-ranking will determine the efficiency of silicon debug.

In this paper, we develop a systematic diagnosis framework to meet the above requirements. We assume that the silicon failures are captured using automated test equipment (ATE) and the circuit-under-test has the DFD capability which can dump internal register values.

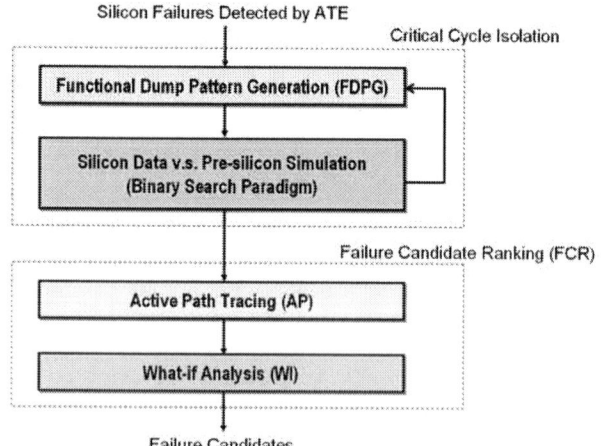

Fig. 1. Silicon failure diagnosis flow.

Figure 1 shows the diagnosis flow of the proposed framework. Given the failures log reported from ATE, we propose a Functional-Dump-Pattern-Generation (FDPG) technique to isolate the critical cycles. We then utilize the failure candidate ranking (FCR) framework which includes

1550-4093/07 $25.00 © 2007 IEEE

active-path-tracing (AP) and what-if analysis (WI). Unlike traditional fault diagnosis approaches, our techniques can perform multiple cycle analysis.

The remainder of the paper is organized as follows. Chapter 2 describes the proposed technique of critical-cycle isolation. Chapter 3 depicts how we perform failure candidate ranking. In Chapter 4, we give a case study and show experimental results. Follows the conclusions and future works.

II. CRITICAL CYCLE ISOLATION

Determining where is the critical cycle is inevitable in functional pattern diagnosis. To effectively isolate critical cycle, firstly we provide the pattern generation feature for ATE which is named FDPG. Next, we iteratively check whether or not the scan-dump data is identical to the pre-silicon simulation. Based on reported information, we can narrow down the critical-cycles for further analysis.

A. Functional Dump Pattern Generation (FDPG)

Dumping register values for a given cycle is still difficult even with DFD capability. Generally, engineers need to modify the test-bench to control the DFD structure at a specific time. Moreover, engineers have to parse the structure of DFD such that the scan-dump data can be mapped back to the design for further analysis.

To relieve DFD to obtain an exact snapshot in a functional test pattern, we provide FDPG. The key idea of FDPG is to analyze the pre-silicon simulation dump file (e.g. VCD file) instead of analyzing the test-bench. Given a simulation dump file, as shown in Figure 2, FDPG will truncate the patterns after the break-point. Then it translates the format of simulation dump file into the imported format of ATE. Furthermore, it appends the necessary control patterns to carry out DFD. Port mapping, consistency checking, and other setup for the ATE are also generated. In the end, the translated pattern can let DFD exactly dump internal states for an original functional test pattern.

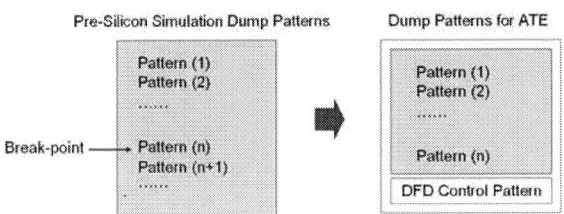

Fig. 2. Translating pre-silicon simulation into dump patterns for ATE.

B. Binary Search Paradigm

Now that we are able to compare the silicon data with the simulation values at a specific cycle, we can apply binary search paradigm to isolate the critical cycle. Take Figure 3 for example. Assume the first failed primary outputs occur at cycle 1000. Then binary search suggests that we need to dump and compare silicon data at cycle 500 firstly. If no mismatches happen at cycle 500, we reason that the first problematic cycle

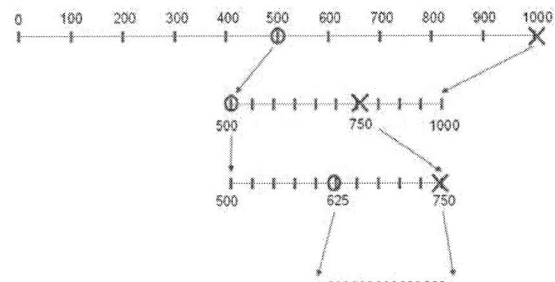

Fig. 3. Example of binary search paradigm for isolating critical cycle.

must locate between cycle 501 and cycle 1000. Otherwise, the critical cycle is between cycle 0 and cycle 500. In a word, given a silicon failure for primary outputs at cycle n, we can use DFD with FDPG and apply binary search paradigm to isolate the critical cycle within log(n) times.

III. FAILURE CANDIDATE RANKING (FCR)

The goal of failure candidate ranking (FCR) is to identify some possible signals that behave different between silicon data and pre-silicon simulation. Furthermore, it must give scores for each candidate to suggest the possibility of being the actual root-cause. In general, FCR takes the circuit gate-level netlist, the golden pre-silicon simulation dump file, and the faulty scenario obtained in critical-cycle isolation as inputs. Figure 4 shows an example of a faulty scenario. Simply put, a faulty scenario must contain three parts: (1) c_T: the correct cycle time in the simulation. That is, no signal mismatches can happen at c_T. (2) m_T: the first mismatched cycle time in the simulation; and (3) a list of the mismatched signals at m_T. By using this faulty information, the following two techniques can reason circuit logic and root cause the failures.

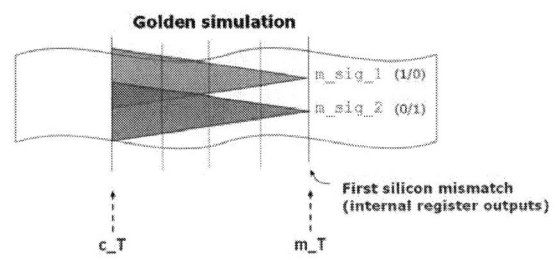

Fig. 4. The faulty scenario for failure candidate ranking (FCR).

A. Active Path Tracing (AP)

Starting from a mismatched signal at m_T, active path tracing (AP) back-traces the circuit structure and also identifies the active fan-in signals. For a gate, an input signal is active means when that signal has a value change, the output of the gate changes value correspondingly. In other words, if we see that the output of a gate mismatches the expected value, then the inputs that are active in the golden simulation are likely to be the failure candidates. Figure 5 shows the active conditions for an AND gate. Of course, determining whether an input of a

1550-4093/07 $25.00 © 2007 IEEE   95

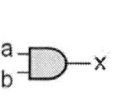

a	b	x	Active Fan-in
0	0	0	a, b
0	1	0	a
1	0	0	b
1	1	1	a, b

Fig. 5. Active fan-in conditions for an AND gate.

gate is active depends on the relationship of the input-output values.

Based on the faulty scenario shown in Figure 4, the internal cycles between c_T and m_T may be more than one. Moreover, multiple clock domain design is very popular nowadays. Therefore, AP also needs to analyze the sequential elements and examine the operation of the clocking information for these elements. In golden pre-silicon simulation, such information is available and thus AP can handle them to precisely locate which signal at which time may cause failures.

In our framework, we assume "single-failure-at-one-time" when performing failure diagnosis [6]. Since debugging is an iterative process, such strategy can easily implicate the components of complicated failure behaviors. Under this assumption, the intersection of the active candidates of different mismatched signals will be given higher scores. Based on the ranking of scores, we cut off the less possible candidates and then execute the "what-if" analysis to further evaluate the remaining suspects.

B. What-if Analysis (WI)

What-if (WI) analysis provides the capability of injecting a faulty value for a failure candidate at a specified time. Then a faulty simulation is performed and the responses of the candidate's fanout signals at m_T are observed. Figure 6 depicts the manner of WI, where the signal c is the candidate for evaluation, the signals $\{m\_sig\_1, m\_sig\_2\}$ represents the fanout mismatched signals to be observed, and the signals $\{c\_sig\_1, c\_sig\_2\}$ represents the fanout correct signals to be observed at m_T. Conceptually, WI is similar to the force/release action in Verilog simulation. If the values of the observed signals match the silicon data, then such candidate is most likely to be the root cause of the failure.

While executing WI for a failure candidate, the behavior of the observed signals can be classified into three categories, which is similar to the POIROT package [3].

(1) Explained: the value of an observed signal matches the silicon data.

(2) Not-explained: the silicon data is correct while the value of an observed signal is incorrect.

(3) Mis-prediction: the silicon data is incorrect while the value of an observed signal is correct.

We apply different weights for the three categories (heuristically, (1) > (3) > (2)), and we sum up the total weight to determine the score for that candidate. For example, assume a failure candidate x_1 has three fanout signals at m_T: y_1, y_2, and

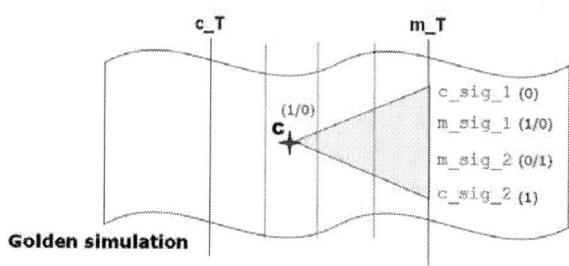

Fig. 6. What-if analysis: inject a faulty value and evaluate the responses.

y_3. After WI, we obtain y_1 is explained, y_2 is mis-prediction, and y_3 is not-explained. Thus, the score of x_1 will be $a_1*y_1 + a_2*y_2 + a_3*y_3$, where a_1, a_2, and a_3 is the weight for each category. In the end, we can rank the failure candidates by the scores to suggest engineers to confirm physical failure exploration.

IV. CASE STUDY

We carried out a case study for an industrial benchmark to evaluate the proposed diagnostic framework. The benchmark is a 330K gate-count, 667 primary outputs sub-system, where about 30K sequential elements are included. Two internal clock domains are also found in the circuit. We assume the benchmark has DfD so that we can see the values of all registers at a specific cycle.

We injected two faults, *fault-1* at cycle 55 and *fault-2* at cycle 1305, to demonstrate the contribution of our approach. The length of a cycle is determined by the ATE; thus, one cycle may contain more than one internal clock period of the circuit. We assume both faults only last within one cycle. As a result, the failure for the injected faults was captured by the primary outputs at cycle 1317.

We attempted to isolate the critical cycle between cycle 0 and cycle 1317 in binary search manner. Finally, we learned that cycle 56 is the critical cycle, where only one register value mismatch was found. Based on the information of the critical cycle, we can derive the faulty scenario for FCR: (1) $c\_T =$ cycle 55, (2) $m\_T =$ cycle 56, and (3) one mismatched signal at m_T. Figure 7 sketches such scenario.

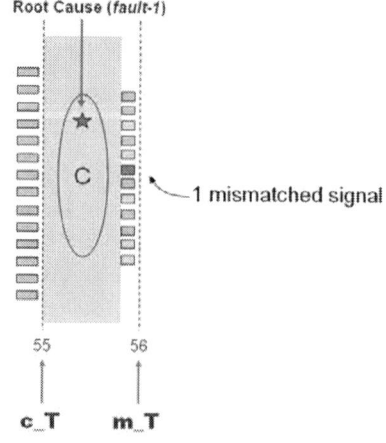

Fig. 7. Faulty scenario for diagnosing *fault-1*.

1550-4093/07 $25.00 © 2007 IEEE

After performing FCR on the faulty scenario shown in Figure 7, we obtained only one failure candidate that is just the *fault-1*. To confirm whether *fault-1* at cycle 55 can cause failures for the primary outputs at cycle 1317, we carried out the "what-if" analysis for *fault-1* and observed the responses for primary outputs from cycle 56 to cycle 1317. Unfortunately, no failures occurred due to *fault-1*. In other words, *fault-1* at cycle 55 cannot contribute any effects to the failures at primary outputs. We thus concluded that *fault-1* is a redundant fault.

Now we needed to go back the step of critical cycle isolation. Furthermore, since we had known that *fault-1* is redundant, we must neglect the mismatched effect caused by *fault-1*. As a result, we obtained cycle 1306 is the critical cycle, where 17 mismatches were found. The faulty scenario was also derived and shown in Figure 8.

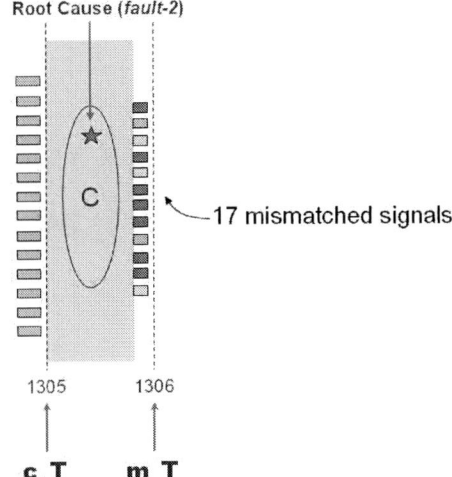

Fig. 8. Faulty scenario for diagnosing *fault-2*.

FCR on the faulty scenario shown in Figure 8 gave 20 failure candidates in the first ranking, including *fault-2*. Similarly, we used "what-if" analysis for those candidates to double check the responses of primary outputs between cycle 1306 and cycle 1317. However, the faulty effect of the 20 candidates cannot be distinguished. This is due to the limitation of only one functional test pattern. To enhance the diagnostic resolution, more functional test patterns or user knowledge are needed.

In practice, sometimes the isolated critical cycle may be larger than one cycle, which depends on the capability of ATE and the feature of DFD. To handle such situations, we set up four different faulty scenarios on the benchmark and applied FCR to them. Table I summarizes the faulty scenarios and also gives the experimental results, where we have discussed the *scenario-1* above. The data shown in Table I was operated under AMD Opteron 252 processor with 8MB memory. We set the run time limit as 10000 seconds and go to terminate FCR when exceeding the run time limit.

In Table I, the 2nd, 3rd, and 4th column represents the faulty scenario c_T, m_T, and the number of mismatched signals at m_T respectively. Column 5 shows the number of failure candidates in the first ranking after performing FCR, and

TABLE I.
EXPERIMENTAL RESULTS ON DIFFERENT SCENARIOS.

	Faulty Scenario for FCR			FCR Results		
	c_T	m_T	# mismatched signals	# 1st Failure Candidates	Time	Memory
scenario-1	1305	1306	17	20	592s	1.4G
scenario-2	1305	1307	276	22	1470s	1.4G
scenario-3	1305	1306	766	22	6909s	1.6G
scenario-4	1305	1309	801	N/A	>10000s	1.4G

column 6 and column 7 shows the run time and memory usage. From the *scenario-1* to *scenario-3*, actual root cause, *fault-2*, is included among the candidates of the first ranking. Based on the result, we demonstrate that FCR can get promising results in diagnosing functional test patterns, even across many cycles. The *scenario-4* cannot be finished due to run time limit, and how to improve the performance of FCR will be our future work.

V. CONCLUSIONS AND FUTURE WORKS

We develop a novel framework to diagnose the root causes for the silicon failures in functional test patterns. The framework can systematically isolate the critical-cycle by utilizing the DFD feature in the circuit. Furthermore, we propose FCR to automatically root-cause the failure candidates. FCR contributes sequential diagnosis capability, which can take clock information into consideration. Experimental results show that our framework is an effective way to perform silicon failure diagnosis.

Future works will target at the following investigation:
· Improve performance for FCR
· Enhance FCR resolution by incorporating other circuit information (e.g. timing)
· Handle situations that DFD can only dump partial register values
· Handle situations that pre-silicon simulation is not available

REFERENCES

[1] B. Vermeulen and S. K. Goel, "Design for Debug: Catching Design Errors in Digital Chips", IEEE Design & Test of Computers, May-June 2002.

[2] M. Abramovici, P. Bradley, K. Dwarakanath, P. Levin, G. Memmi, and D. Miller, "A Reconfigurable Design-for-Debug Infrastructure for SoCs", *Proc. IEEE/ACM Design Automation Conf. (DAC)*, 2006.

[3] S. Venkataraman and S. B. Drummonds, "Poirot: Applications of a Logic Fault Diagnosis Tool", *IEEE Design & Test of Computers*, Jan.-Feb. 2001.

[4] P. Y. Chung, Y. M. Wang, and I. N. Hajj, "Diagnosis and Correction of Logic Design Errors in Digital Circuits", *Proc. IEEE/ACM Design Automation Conf. (DAC)*, 1993.

[5] S-Y. Huang K-T. Cheng, K-C. Chen, and D-T. Cheng, "ErrorTracer: A Fault Simulation Based Approach to Design Error Diagnosis", *Proc. Int'l Test Conf. (ITC)*, 1997.

[6] T. Barternstein, D. Heaberlin, L. Huisman, and D. Sliwinski, "Diagnosing Combinational Logic Designs Using the Single Location At-a-Time (SLAT) Paradigm", *Proc. Int'l Test Conf. (ITC)*, 2001.

[7] M. Paniccia, T. Eiles, V. R. M. Rao, and W-M. Yee, "Novel Optical Probing Techniques for Flip-Chip Packaged Microprocessors", *Proc. Int'l Test Conf. (ITC)*, 1998.

1550-4093/07 $25.00 © 2007 IEEE

AUTHOR INDEX

Abadir, Magdy 33
Al-Asaad, Hussain 9
Al-Sukhni, Hassan 61
Arons, Tamarah 45
Bamford, Noah 52
Bangalore, Rekha K 52
Becker, Bernd 37
Bernardi, P. 3
Bhadra, Jayanta 33
Bhan, Deepa 15
Bolzani, L. 3
Caldwell, James K. 15
Campos, Jorge 9
Chapman, Eric 52
Chavez, Hector 52
Dasari, Rajeev 52
Elster, Elad 45
Gangaram, Vijay 15
Herbstritt, Marc 37
Holt, James 61
Hsu, Yu-Chin 94
Jimenez, Edgar 52
Kim, Hyun Sung 76
Koo, Heon-Mo 33
Li, Sikun 83

Lin, Hermes 94
Lin, Ten 94
Lin, Yinfang 52
Lindberg, David 61
Liu, Tayung 94
Manzone, A. 3
Mishra, Prabhat 33
Murphy, Terry 45
Osella, M. 3
Patel, Hiren D. 68
Reese, Michele 61
Reorda, M. Sonza 3
Safarpour, Sean 88
Scholl, Christoph 37
Shukla, Sandeep K. 68
Singerman, Eli 45
Sziray, József 20
Veneris, Andreas 88
Violante, M. 3
Walker, D. M. H. 76
Yan, Ming 83
Yang, Kai 94
Yen, Chia-Chih 94
Zhang, Jianmin 83

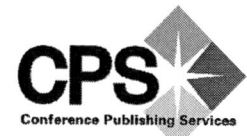

IEEE Computer Society
Conference Publications
Operations Committee

CPOC Chair
Phillip Laplante
Professor, Penn State University

Board Members
Mike Hinchey, *Director, Software Engineering Lab, NASA Goddard*
Linda Shafer, *Professor Emeritus, University of Texas at Austin*
Jeffrey Voas, *Director, Systems Assurance Technologies, SAIC*
Thomas Baldwin, *Manager, Conference Publishing Services* (CPS)

IEEE Computer Society Executive Staff
David Hennage, *Executive Director*
Angela Burgess, *Publisher*

IEEE Computer Society Publications
The world-renowned IEEE Computer Society publishes, promotes, and distributes a wide variety of authoritative computer science and engineering texts. These books are available from most retail outlets. Visit the CS Store at *http://www.computer.org/portal/site/store/index.jsp* for a list of products.

IEEE Computer Society *Conference Publishing Services* (CPS)
The IEEE Computer Society produces conference publications for more than 200 acclaimed international conferences each year in a variety of formats, including books, CD-ROMs, USB Drives, and on-line publications. For information about the IEEE Computer Society's *Conference Publishing Services* (CPS), please e-mail: tbaldwin@computer or telephone +1-714-821-8380. Fax +1-714-761-1784. Additional information about the IEEE Computer Society's *Conference Publishing Services* (CPS) can be accessed from our web site at: *http://www.computer.org/cps*.

IEEE Computer Society / Wiley Partnership
The IEEE Computer Society and Wiley partnership allows the CS Press *Authored Book* program to produce a number of exciting new titles in areas of computer science and engineering with a special focus on software engineering. IEEE Computer Society members continue to receive a 15% discount on these titles when purchased through Wiley or at: *http://wiley.com/ieeecs*. To submit questions about the program or send proposals, please e-mail dplummer@computer.org or telephone +1-714-821-8380. Additional information regarding the Computer Society's authored book program can also be accessed from our web site at: *http://www.computer.org/portal/pages/ieeecs/publications/books/about.html*.

Revised: 17 August 2006

We're proud to announce the launch of *CPS Online*, a new IEEE online collaborative conference publishing environment designed to speed the delivery of price quotations and provide conferences with anytime access to all of a project's publication materials during production, including the final papers. **CPS Online**'s workspace gives a conference the opportunity to upload files through any Web browser, check status and scheduling on a project, make changes to the Table of Contents and Front Matter, approve editorial changes and proofs, and communicate with a CPS editor through discussion forums, chat tools, commenting tools and e-mail.

The following is the URL link to the CPS Online Publishing Inquiry Form:
http://www.ieeeconfpublishing.org/cpir/inquiry/cps_inquiry.html

9780769528397